Postdigital Storytelling

Postdigital Storytelling offers a groundbreaking re-evaluation of one of the most dynamic and innovative areas of creativity today: digital storytelling. Central to this reassessment is the emergence of metamodernism as our dominant cultural condition.

This volume argues that metamodernism has brought with it a new kind of creative modality in which the divide between the digital and non-digital is no longer binary and oppositional. Jordan explores the emerging poetics of this inherently transmedial and hybridic postdigital condition through a detailed analysis of hypertextual, locative mobile and collaborative storytelling. With a focus on twenty-first century storytelling, including print-based and non-digital art forms, the book ultimately widens our understanding of the modes and forms of metamodernist creativity.

Postdigital Storytelling is of value to anyone engaged in creative writing within the arts and humanities. This includes scholars, students and practitioners of both physical and digital texts as well as those engaged in interdisciplinary practice-based research in which storytelling remains a primary approach.

Spencer Jordan is Assistant Professor of Creative Writing in the School of English at the University of Nottingham, UK. He is both a published novelist and historian, with a background in the digital humanities. He has been involved in a number of digital projects exploring immersive and location-based storytelling. Particular areas of interest include: digital/hypertext and immersive fiction, literary geography and digital heritage.

Digital Research in the Arts and Humanities

Series Editors: Marilyn Deegan, Lorna Hughes, Andrew Prescott, Harold Short and Ray Siemens

Digital technologies are increasingly important to arts and humanities research, expanding the horizons of research methods in all aspects of data capture, investigation, analysis, modelling, presentation and dissemination. This important series covers a wide range of disciplines with each volume focusing on a particular area, identifying the ways in which technology impacts on specific subjects. The aim is to provide an authoritative reflection of the 'state of the art' in technology-enhanced research methods. The series is critical reading for those already engaged in the digital humanities, and of wider interest to all arts and humanities scholars.

The Shape of Data in the Digital Humanities
Modeling Texts and Text-based Resources
Julia Flanders and Fotis Jannidis

International Perspectives on Publishing Platforms
Image, Object, Text
Edited by Meghan Forbes

A History of Place in the Digital Age
Stuart Dunn

The Historical Web and Digital Humanities
The Case of National Web Domains
Edited by Niels Brügger and Ditte Laursen

Postdigital Storytelling
Poetics, Praxis, Research
Spencer Jordan

To learn more about this series please visit: www.routledge.com/Digital-Research-in-the-Arts-and-Humanities/book-series/DRAH

Postdigital Storytelling

Poetics, Praxis, Research

Spencer Jordan

LONDON AND NEW YORK

First published 2020
by Routledge
2 Park Square, Milton Park, Abingdon, Oxon OX14 4RN

and by Routledge
605 Third Avenue, New York, NY 10017

First issued in paperback 2021

Routledge is an imprint of the Taylor & Francis Group, an informa business

British Library Cataloguing-in-Publication Data
A catalogue record for this book is available from the British Library

Library of Congress Cataloging-in-Publication Data
Names: Jordan, Spencer, author.
Title: Postdigital storytelling : poetics, praxis, research / Spencer Jordan.
Description: London ; New York : Routledge, 2019. |
Series: Digital research in the arts and humanities |
Includes bibliographical references and index.
Identifiers: LCCN 2019031775 (print) | LCCN 2019031776 (ebook) |
ISBN 9781138083509 (hardback) | ISBN 9781315112251 (ebook) |
ISBN 9781351621472 (epub) | ISBN 9781351621465 (mobi) |
ISBN 9781351621489 (adobe pdf)
Subjects: LCSH: Narration (Rhetoric) | Digital storytelling. |
Creation (Literary, artistic, etc.)
Classification: LCC P96.N35 J67 2019 (print) |
LCC P96.N35 (ebook) | DDC 808/.036–dc23
LC record available at https://lccn.loc.gov/2019031775
LC ebook record available at https://lccn.loc.gov/2019031776

Typeset in Times New Roman
by Newgen Publishing UK

ISBN 13: 978-1-03-208770-2 (pbk)
ISBN 13: 978-1-138-08350-9 (hbk)

Printed in the United Kingdom
by Henry Ling Limited

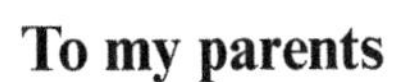

To my parents

Contents

Figures

Acknowledgements

Kind and generous thanks are due to the following for the use of images: Barbara Kinney; Mark Z. Danielewski; Simon Starling; Trevor Paglen; Cornelia Parker; Vera Frenkel; Michael Joyce; Lance Olsen; Scott Rettberg; William Gillespie; Dirk Stratton; Tom Abba; Duncan Speakman; Barrett Lyon; Metro Pictures, New York; Frith Street Gallery, London; The Modern Institute, Glasgow; Turner Galleries, Perth; Tate Images; the Richard Giblett Estate; Murray White Room, Melbourne; John Hawkins & Associates, Inc.; Picture Sunshine Coast; and Mary Evans Picture Library.

I would like to thank Professor Katharine Cox for all her love and support; the School of English, University of Nottingham; Professor Gareth Loudon (Centre for Creativity Ltd. and Cardiff Metropolitan University); Dr. Mike Reddy (the University of South Wales); Dr. Kate Watson; Francesca James (Fresh Content); Cardiff School of Education & Social Policy, Cardiff Metropolitan University; colleagues past and present at the Department of Arts and Cultural Industries, University of the West of England, Bristol; Heidi Lowther, Elizabeth Risch and Jack Boothroyd at Routledge; and the many inspirational students I have had the good fortune to teach.

1 Introduction

Fake news, folks. Fake news.

At the heart of this book is a radical re-evaluation of one of the most dynamic and innovative areas of creativity today, namely digital storytelling. Central to this reassessment is the emergence of a new kind of creative modality in which the divide between the digital and non-digital is no longer binary and oppositional. I term this condition *postdigital*. Although digital storytelling is a key focus, the foundation of this book remains, what I identity as, new and evolving forms of postdigital interconnectedness. As such, the book does not artificially separate digital from non-digital storytelling. Rather it seeks to explore and understand the emerging poetics of a postdigital condition that is inherently transmedial and hybridic, and in which the digital and the non-digital domains are increasingly entangled (Berry and Dieter 2015). Yet this entanglement is not without hierarchy; indeed, this book will show that the recourse to print is increasingly a strategic, even transgressive, act. In the postdigital age, it is the printed book that has become a signifier of functional deficit, a physical manifestation of 'that which is not digital' rather than any natural or intrinsic form of published work in and of itself.

By exploring both the affordance and affectivity of these poetics within the arts and humanities, this book also offers a significant re-evaluation of practice-based inquiry as a critical form of knowledge production within theoretical and research paradigms.

The book's critical perspective is therefore wider and deeper than much of the previous work in this area. Crucially I place the discussion of contemporary digital storytelling within a more profound understanding of *creativity* itself. I argue that, with the end of postmodernity, we have entered a new ontological paradigm, an alternative way of seeing and understanding the world that has, at its heart, a different creative modality. While postmodernity is associated with irony and depthlessness, denying any sense of meaningful representation and action, this new age, what some have called metamodernism (van den Akker *et al.* 2017), embraces innovative forms of situated embodiedness. Critically, too, is a new ethical imperative from which comes a belief in, and desire for, affective action and change in response to social, cultural and ecological crises. In this new creative paradigm, then, creativity becomes a tactical intervention whose end goal is affective change, what Donna J. Haraway calls 'troublemaking' (2016). One of the core arguments of this book is that postdigitality is a fundamental feature of this metamodernist condition and not simply a product of technological advancement. As a consequence, postdigitality emerges from these chapters as a key marker of deep and sustained cultural and social change.

There are two significant consequences of these approaches for my argument: the first is that I necessarily push and extend the definitional boundaries of *digital storytelling*, an umbrella term that, over the years, has come to comprise a wide range of subgenres and tropes, from electronic literature to hypertext fictions. The critical perspectives taken in this book allows these rather tired concepts to be dynamically reconfigured. Here, postdigital storytelling embraces both formal and informal modes and forms, including social media and web-based platforms and apps; it embraces digital biography and non-fiction as much as fiction, prose and poetry. It is code, data, narrative and performance. It is collaborative and participatory as well as individual and personal. Yet it is also transmedial, foregrounding the postdigital mash-up where the divide between the digital and non-digital is porous and creatively fluid. As such, one of the innovations of this book is that it examines the response of traditional printed texts to the postdigital condition.

The second consequence relates to the relationship between postdigital storytelling and academic research. With these new configurations and modalities come innovative ways of thinking about the role and function of creativity as both praxis and research (Barrett and Bolt 2010). Specifically I argue how new interdisciplinary and transdisciplinary forms and approaches are needed, rethinking the traditional division between creative artefact and critical exegesis. In fact, I demonstrate how postdigital storytelling and transdisciplinarity arc irrevocably intertwined, each the child of our new ontological paradigm. Storytelling has much to offer the research of global imperatives such as resilience and empathy (Bazalgette 2017), yet it is only by reaching across and beyond academic boundaries that such issues will be addressed (see Koehler 2017). And that reaching across comes not just from the arts and humanities disciplines. In the interdisciplinary terrain of science and technology studies, for example, Ulrike Felt reminds us that concepts such as situated knowledge production and performativity, 'values, aspirations, and imaginaries' (2017, 253), are equally important to those working within science-based disciplines.

The intended audience for this book is therefore threefold. First are authors, writers and theorists of stories *per se*. Since this book embraces the full panoply of output, from printed text, to completely digital work, with all manner of hybridic work in between, the book offers an opportunity to explore continuities and connections across creative modes that have traditionally been remote from each other. Second, are those interested in the wider role of contemporary creativity within a theorisation of postdigital normativity; and third, are those with a concern for practice-based research, particularly the degree to which new and innovative forms of storytelling can inform our understanding of interdisciplinary and transdisciplinary research within the arts and humanities.

Two wider phenomena form the context for this study. The first is the transformation of storytelling since the advent of the new millennium, driven by an ever-growing list of technological innovation. Chief among them are two interrelated global phenomena: the rise of smartphones and the growing ubiquity of social media platforms. These two things alone have radically altered how stories are both constructed and consumed. From being static, PC/CD-ROM dependent artefacts, the traditional digital story has been freed from its gilded cage. The adjective 'hypertext' has found itself replaced by an ever increasing list of arriviste upstarts (including, but certainly not limited to, shareable, locative, adaptive, mashable, generative, augmented, virtual and collaborative) as digital stories have been transformed by a new generation of functionality.

Alongside these recent technological developments, there has also been a concomitant rise in the perceived value and utility of storytelling *per se* within the arts and humanities. As I've intimated already, this is not specific to this discipline area, of course: as Boyd has made clear, stories play a fundamental role within human cognition more generally. In science and technology studies, for example, an exploration of how knowledge is (co) constructed, shared and challenged remains at the core of the field (Felt 2017) while in health sciences, storytelling has been used with great effect, particularly in the diagnosis, treatment and research of mental health conditions such as depression, anxiety and trauma, but also in the way clinicians share and understand their own reflective practice (for example, see Gabbay and le May 2011).

Traditionally, the value of practice-based approaches (such as storytelling) across the arts and humanities was itself rather neglected, denigrated to the methodological sidelines of those disciplines perceived to come within the remit of creative art and performance as opposed to the humanities (Barrett and Bolt 2010). Yet a brief overview of humanities disciplines will show that, over the last ten years, things have begun to change quite radically. As the *Digital Research in the Arts and Humanities* series is showing, disciplines as disparate as geography, archaeology, literature, linguistics and history have all, to some extent or other, begun to explore the potential of storytelling as a practice-based research methodology (see Smith and Dean 2009). What have been termed the 'spatial' (Cooper *et al.* 2017), 'creative' (Harris 2014) and 'participatory' (Facer and Enwright 2016) turns are transforming research across the humanities (and beyond), and this book will show how storytelling is an important element of each. As Estelle Barrett notes, 'practice-led research is a new species of research, generative enquiry that draws on subjective, interdisciplinary and emergent methodologies that have the potential to extend the frontiers of research' (2010, 1). It is hoped that this book goes at least some way in demonstrating the veracity of Barrett's assertion.

It is these two conditions then, the growing affordance offered by digital platforms for storytelling, *and* the wider interest from arts and humanities disciplines in the value of practice-based approaches to knowledge creation, that inform the context of this book. By seeking to map this new and emergent terrain, the book is innovative and timely, both for the study of postdigital storytelling as a new creative modality, but also through an understanding of its specific impact on the arts and humanities disciplines.

Exploring the role of storytelling within the wider context of practice-led research will be key here. As Hazel Smith and Roger T. Dean note (2009), such research needs to be understood as an essentially bi-directional process. In other words, practice-led research is as much about the production of creative output that leads to research, as it is about research in itself that then leads to creative practice. Smith and Dean call this an 'iterative cyclic web' and it forms one of the cornerstones of this work. Crucially, what I'll be developing here will build on and extend Barrett's 'interdisciplinary' methodology (in other words, research that crosses academic boundaries). Instead, and as I've already intimated, I shall be using postdigital storytelling as a critical focus through which to advance a more transformative framework for *transdisciplinary* research across and beyond the arts and humanities. In other words, the book seeks to prioritise research and methodological approaches that bring together both scholarly disciplines and non-academic stakeholders to explore what Jay Bernstein, in his study of transdisciplinarity, calls 'the inherent complexity of reality' (2015, 13).

Still life: politicians with smartphones

In September 2016, Hillary Clinton, the Democratic candidate in that year's American presidential election, emerged into a room full of supporters. What happened next produced one of the most controversial photographs of the campaign in which, according to Chris Graham of *The Daily Telegraph*, a crowd of 'narcissistic' millennials revealed their desperation to take a photograph with their political heroine (2016). In fact, as the person who took the photograph, Barbara Kinney, later made clear, the mass selfie had been Clinton's idea.

The photograph (as shown in Figure 1.1) is striking and somewhere within it is surely a message for our times. That message, however, is not about narcissism, at least not directly. Instead, what the photograph really shows, better than countless pages of statistics and graphs, is something that is all too easy to overlook today, namely both the speed and depth of the change to our everyday behaviour brought about by digital technology. Imagine going back in time, say to 2006, and presenting this same photograph to a group of tech-savvy young adults. They wouldn't have a clue what was going on; go back even further, to 1996, and the bemusement would be even greater. Such confusion would not simply be created by all those strange silver objects in the crowds' hands. It would also be engendered by the physical actions of the people themselves, that strange act of mass cold-shouldering, arms raised, as though not daring to lay eyes on what has emerged before them, and, like Perseus and his shield, gaze only at a reflection caught by whatever they are holding. There's a word for that action of course. We know it but all of those living in our thought experiment wouldn't have done. The term selfie only really took off in 2010 with the arrival of the front-facing camera on smartphones such as the iPhone 4 (Losse 2013). Yet its cultural impact has been precipitous. By 2013 the Oxford English Dictionary had christened selfie as their word of the year, having beaten such shortlisted contenders as bedroom tax, bitcoin and twerk. It is this speed of normalisation and technical acculturation that is captured by the photograph of Hillary Clinton. In this way we can see that the photograph acts as a kind of synecdoche of the wider technical and cultural changes that have transformed society since the advent of the new millennium. The selfie and the ritualised communal behaviour of its taking are the more visible reminders of just how fundamental has been the impact of digital technology on our lives.

One area where this influence continues to be acutely felt is in the area of storytelling. Storytelling of course could be considered to be a rather archaic term in itself, conjuring up images of night-time ghost stories told orally around flickering flames and dancing shadows. From an academic perspective, narrative or (hyper) textual construction might seem a better choice and certainly narrative and text are preferred by some theorists, particularly those approaching digital texts from a linguistic perspective (for example, see Bell 2010; Ensslin 2014; Ryan 2015). Yet storytelling captures something that perhaps is lost with other terms. For a start, the term neatly encapsulates two separate, though intertwined, elements: both the creation of a story and its telling. As a verb, storytelling helps to represent the underlying iterative process between making and telling, between the act of creating a story, and the separate, though interrelated, act of providing access to that story for an audience. Storytelling therefore prioritises both author and media (paper, oral, digital) as much as it does the reader. And, as we'll see, decisions about, what might be termed, means of access, directly influence both the nature and form of the narrative. In other words, the process by which an audience accesses a story is as important as the story itself. In fact, the shift away from the traditional oral and paper-based forms of transmission, towards digital platforms and forms, only extenuates this phenomenon. The functionality offered by digital technology inevitably leads to a radical reassessment of both form and structure, perhaps best exemplified by the transformation of print-based journalism and the media sector (Gauntlett 2015). As John Naughton noted of the internet more generally, disruption 'is a feature of the system, not a bug' (2012, 4–5). As we'll see, storytelling has certainly not been immune to its own form of disruption. In fact, the inherent transformative capacity of digital technology remains at the heart of this book.

Yet there is another reason why storytelling should be considered an apposite term. Since the dawn of the new millennium, there has been an increasing social and cultural imperative to understand what might be, rather grandly, called the twenty-first century human condition. Although incredibly broad, it is possible to identify two key characteristics of this imperative. The first would be the growing recognition of the value of transdisciplinary activity and approaches already highlighted; in other words, research that transcends traditional subject-based boundaries. The second characteristic would be a focus on 'community' as a critical site of social and cultural value (Facer and Pahl 2017). This is important as it marks an epistemological shift away from conceptual frameworks that favour unproblematic notions of 'society', 'city' and 'nation'. A single place will have multiple, conflicting, communities, defined by a wide, and ever changing, range of factors, that could, although might not, include such things as background, age, race, gender and sexuality (Lambert 2013). This imperative has been given fresh impetus through the election of the 45th President of the United States of America, Donald Trump, in November 2016, and the EU (Brexit) Referendum in the United Kingdom in June 2016. Both events gave outcomes that dumbfounded many critics and academic experts (Welfens 2017). In the aftermath, there has been a scramble to re-engage with the actualité of life as it's lived on the street, whether it be the rust belt towns and cities of America, or the Brexit citadels of small town Britain (for example, see Raghuram Rajan's *The Third Pillar: The Revival of Community in a Polarised World* 2019). Crucially, many experts have highlighted the importance of social media in the creation of enclosed 'ecosystems' of news, generating hermetically-sealed 'clusters of like-minded people sharing like-minded things' (*The Economist* 2017). Suddenly community-based research has become all about saving the liberal democratic dream. Yet, as Keri Facer and Kate Pahl admit, research that is both interdisciplinary and community-based, is liable to be 'messy, contingent on practice, uncertain, embedded in stories and histories that could be dismissed as "anecdotal", and located in events and histories that are themselves ephemeral and lacking disciplinary anchorage' (2017, 11). If we are to understand communities and the complexities of people's lives then, Facer and Pahl are just two of many academics calling for the prioritisation of storifying as a key research method. Or, to put it another way, it is now recognised that stories play a central role in any understanding of how individuals, communities and societies engage with and understand the world. Storifying then is no longer an activity that is seen as the preserve of a literate, articulate elite. Instead, the basic urge to storify, to generate explanatory narratives in the here and now and on the go, individually within our own minds, and mutually, across dynamic networks and relationships, both real and virtual, should be understood as a fundamental aspect of twenty-first century human cognition.

It is no surprise to find that it is in arts and humanities subjects where there has been the most generosity shown towards these ideas. Such endeavours are very often interdisciplinary in nature, embracing what has come to be called the 'linguistic turn' in methodological approach. Building on the work of theorists such as Michel Foucault (1986) and Henri Lefebvre (1991), more recent work has focussed on the importance of place and space as key concepts, leading to, perhaps rather inevitably, the adoption of the term spatial turn. This kind of approach foregrounds the performative and spatial aspect of knowledge creation, in which storytelling plays a key role. This in turn naturally leads to radical ideas of cultural mapping and topographic representation, what Les Roberts calls the 'textualities of space, place and mapping' (2015, 18). Lawrence Cassidy's 'cultural memory project' (2015), for example, is a good example of this approach in which the 'lost history' of two working-class communities, one in Salford, UK, the other in Cape Town, South Africa, are brought together through 'participatory mapping' into a series of interactive art installations consisting of material remnants and fragments.

Yet this interest in storytelling extends much further than those subjects traditionally associated with the humanities. Human geographers have long argued for the centrality of narrative in the subjective engagement with physical space (for example, see Malpas 1999). Recently a focus on storytelling and creative writing has become much more noticeable, with a concomitant interest in materiality, affect and embodiment. Angharad Saunders, for example, argues that it is time geographers began to seriously consider 'the role of writing in the creation and validation of geographical knowledge' (2010, 442). She suggests that it is creative writing's very 'ambiguity, evasiveness and subjectivity' that remains its strength as well as its ability to engage with what, Angharad calls, the 'ordinariness' of everyday life (450). This is emphasised by Emilie Cameron who notes that it is creative writing's ability to 'affect' that is one of its most important characteristics: 'the capacity for stories to be practiced in place and to generate (intersubjective) change' (2012, 581).

Stories then help us to understand and explore both the situatedness of meaning and its capacity for change. Even in a subject such as archaeology, a discipline steeped in the scientific lore of artefactual discovery and interpretation, there is ongoing debate about the utility of creative storytelling. For Gavin Lucas, 'archaeology is a materialising activity – it does not simply work with material things, it materialises. It brings new things into the world; it reconfigures the world' (2004, 117). It is in this process of 'materialising' that Lucas sees the role of storytelling, foregrounding the role of human subjectivity in the creation of meaning. Gabriella Giannachi, Nick Kaye and Michael Shanks are even more emphatic. They argue that archaeology is less about understanding the past and more about how narratives of the past are constructed and experienced. From this perspective, they foreground the experience of place through what they call 'the performance and construction of the past in memory, narrative, [and] collections (of textual and material sources)' (2012, 2).

Ruth Tringham's *Dead Women Do Tell Tales* project created a digital database of Neolithic archaeology recovered from excavations in southeast Europe and Anatolia (2015). The choice of electronic platform for the reader allows the creation of, what Tringham calls, 'recombinant histories', made up of narratives that are 'collections of indefinitely retrievable fragments ... by which the past may be conceived as fundamentally mutable and reconfigurable' (Anderson 2011). In terms of narrative, Tringham notes the similarity between these 'recombinant histories', steeped in the intimate and everyday detail of microhistories, and the partiality of flash fiction (28–29). The effect on the reader is the same: stories that deliberately undermine any sense of an objective, readily graspable, truth or understanding. There's something else going on too: the generation of poems and stories from pre-existing textual fragments has a long history within the avant-garde. Yet Marjorie Perloff (2010) argues that digital media has made such intertextuality ubiquitous across social and cultural life, to the point where it is now the key creative *modus operandi*, a process Lev Manovich (2001) calls 'cultural transcoding' where digital technology slowly transfigures the practices and behaviours of the non-digital world. Hayles' (2012) concept of 'technogenesis' takes this even further, seeing both human and technical evolution as intimately interdependent. This suggests that academic work such as Tringham's *Dead Women Do Tell Tales* project and Cassidy's cultural memory projects touch on something far-more significant than perhaps might at first appear. From this perspective, concepts such as 'recombinant history' and 'participatory mapping' embrace a fundamental ontological shift in how we both create and understand stories. And at the heart of this change is the impact of digital technology.

It is these issues, amongst others, that this book seeks to explore. Yet for now it is enough to recognise what has been said so far in this introduction. First, that digital technology has impacted on how stories are both written and experienced; and second, and equally important, that storytelling itself has become mainstreamed as a way of understanding the world and what might be called 'lived experience', a epistemological position where meaning is both individualised as well as community (and spatially) based. When the archaeologist Reinhard Bernbeck states that 'archaeology is ultimately nothing other than a huge machine that churns out reasoned ways to bridge gaps in knowledge' (2015, 266), he could not only be talking about every academic discipline, including those in the hard sciences, but also the way we all live our lives. This book argues that storytelling helps us to bridge those gaps, to link together those certainties we feel we can trust; yet it does something else too – it can point a finger at that very process, at the way knowledge is created, and in that way it can foreground the very uncertainties that lie at the heart of all experience.

The postdigital condition

There's a curious anomaly at the heart of digital storytelling. On the one hand the term appears to position itself in the camp of the avant-garde and the dangerously innovative. David Ciccoricco, for example, describes how digital fiction falls naturally within the traditions of experimental literature, a tradition 'typified by its subversion of conventional form, technique, and genre' (2012, 469). Such iconoclasm has helped pigeon-hole digital storytelling as the essence of postmodern fiction, the sort of storytelling that Brian McHale called 'an illusion-breaking art', with its insistence on the foregrounding of the underlying ontological structure of texts and their fictional worlds (1999, 221). Marie-Laure Ryan is in accordance here, championing what she sees as the 'natural and elective affinities of electronic writing for postmodern aesthetics' (1999, 103). A good example of this would be Shelley Jackson's *Patchwork Girl* (1995), developed using the Storyspace software programme and distributed as a standalone CD-ROM. The story is a rewrite of Mary Shelley's *Frankenstein* (1818) while also drawing on the character of Scraps in *The Patchwork Girl of Oz* by L. Frank Baum (1913). In Shelley's original novel, Victor Frankenstein decides to create a female companion for his monster but in the end decides to destroy it before it can be completed. In *Patchwork Girl*, Jackson brings this female monster to life. The story consists of five sections composed of textual fragments and images connected through multiple links. As the patchwork girl explains right at the very beginning of the story, 'I am buried here. You can resurrect me, but only in piecemeal. If you want to see the whole you will have to sew me together yourself'. *Patchwork Girl* draws on ideas of pastiche, appropriation and self-construction. As Alice Bell notes, the experience is ontologically and epistemologically challenging (2010, 148), as the reader tries to put back together the patchwork girl while navigating multiple branching pathways.

A more recent example would be Joanna Walsh's digital novella, *Seed* (2017). Published by Visual Editions and Google's Creative Lab, *Seed* tells the story of an eighteen-year-old girl over a period of four months in the late 1980s. However, while *Patchwork Girl* is a story locked into the 1990s architecture of the desktop computer, *Seed* has been designed for the smartphone with simple swipe-screen navigation. Again, the story emerges from textual fragments, this time scattered across a network of interlacing 'vines'. The order in which each piece of text is read depends ultimately on the reader's own navigational decisions as they move through the story's rich undergrowth, visually represented by Charlotte Hicks botanical illustrations. As the narrator admits, 'However many times I tell it I still don't know what anything means' (Hawlin 2017). This sense of fracture, of making sense of something which is ultimately beyond us, lies at the centre of a story whose opening line promises 'a story that grows and decays, that becomes entangled and disentangled' (Hawlin 2017). It is 'illusion-breaking art' in the sense that it is the reader who creates the story. As Walsh explains, she wanted readers to 'have no sense of reading left to right, of the weight of the book, of how far they were through, or, sometimes, of direction within the narrative' (Hawlin 2017). This sense of directionlessness harks back to earlier printed works such as B. S. Johnson's *The Unfortunates* (1969), a novel made-up of unbound chapters placed in a box, that, apart from the specified first and last, can be pulled out and read in any order. Yet the effect of digital stories such as *Seed* is even more disconcerting. Excluding those first and last chapters, *The Unfortunates* has twenty-five randomised sections. Yet the reader is always aware of the entirety of the novel, sitting there in its famous box. With a work like *Seed*, however, this is not the case. As Walsh admits, 'readers have told me they're unsure about whether they've reached "the end" of the book … And some readers who have explored the text thoroughly have told me they feel there still might be other sections they might have missed' (McMullan 2017). There's clearly a challenge here for writers in regard to how, and to what extent, they balance readerly freedom against narrative coherence and structure, something we'll be exploring in Chapter 5 (see Punday 2018). Yet for now it's enough to recognise that the traditional notion of a story, of a self-contained explanatory narrative with a clearly defined size and structure, has been left a long way behind within the canon of digital storytelling.

Running alongside this tradition of the avant-garde and the experimental, however, is a more recent condition which, rather than responding to digital technology's novelty and originality, is instead a recognition of its overwhelming presence in everyday life. This is a world in which over 90% of adults in the United Kingdom own a mobile phone, over 70% have a smartphone and 73% use social media (Ofcom 2017). A report by the Economist Intelligence Unit noted that, in the Asia-Pacific region in particular, the smartphone revolution has had a 'profound impact on media and content consumption' (2017, 19). The report goes on to note that it is countries such as South Korea, China, Indonesia and the Philippines who are now at the cutting edge of digital innovation.

Yet the smartphone is just one small part a much wider phenomenon, a phenomenon that has at its roots the increasing ubiquity of digital data flows across the planet. The rise of pervasive connectivity, alongside increasing computer processing power and technological innovation, have overseen a revolution that has overwhelmed almost all aspects of contemporary life. Concepts such as Big Data, the Internet of Things, augmented (AR) and virtual (VR) realities, immersivity and artificial intelligence (AI), are just the latest instalments in this endless sequence of change. And as the photograph of Hillary Clinton shows (Figure 1.1), these changes are not just technological: they reach deep into our very sense of self. Anthropologists Daniel Miller and Heather Horst have gone further, stating that digital technologies are now so pervasive they have actually become 'a constitutive part of what makes us human' (2012, 4). The flipside of that, as Deborah Lupton notes, is that digital technology has become invisible, everywhere but also nowhere (2014, 2).

Florian Cramer, Joseph Tabbi and Mel Alexenberg, amongst others, recognise this new era as postdigital.[1] As we'll see, although definitions of the term vary, at its heart is the sense that we have entered a new kind of relationship with technology. As Tabbi notes, we are now living at a time when not only has most literary work been digitised but also that 'nearly all new writing is now

Figure 1.1 Hillary Clinton poses for selfies, Orlando, Florida, 21 September 2016.
Source: © Barbara Kinney/Hillary for America.

done digitally' (2018, 5). The postdigital world then is not one that has moved *beyond* the digital; instead it is a place where the digital is now ubiquitous and consequently invisible. As Cramer argues, 'in a post-digital age, the question of whether or not something is digital is no longer really important—just as the ubiquity of print, soon after Gutenberg, rendered obsolete all debates (besides historical ones) about the "print revolution"' (2012, 162).

By the end of the first decade of the twenty-first century, the rise of handheld devices (Apple's first smartphone was released in 2007) and ubiquitous wifi connectivity facilitated an explosion in Web 2.0 based applications. The best example of this is provided by the phenomenal success of social media platforms, such as Facebook (officially launched in 2004), Baidu Tieba (2003) and Twitter (2006), and more recently Snapchat (2011) and WeChat (2011). Alongside this, dedicated blogging sites appeared, such as Wordpress (2003) and Tumblr (2007); and digital story platforms such as StorySpace (1987), Twine (2009) and Genarrator (2009). The fate of Storify is interesting in this context. Set up in 2010 as a means of creating stories by embedding social media posts within a narrative, the service quickly found that other social networks started to offer similar functionality, such as Facebook's moments and Twitter's threading. Storify closed in 2018.

The concurrent rise of the e-reader confounded many media specialists who predicted that online reading, especially of longer forms of fiction such as novels, would just not take off; at the same time, initiatives such as *Editions At Play*, a collaboration between Visual Editions and Google's Creative Lab, is showing that there is genuine interest in re-thinking what online narratives might actually be. Most of the stories published through *Editions at Play* have been produced for the smartphone. Their rallying cry is for a new type of book, one written specially for the web, that could be 'data-led, locative, generative, algorithmic, sensor-based, fluid, non-linear, expandable, cookie-ish, personalised, proximal, augmented, real-time, time-sensitive, adaptive, collaborative, and share-y' (*Editions at Play*). *Breathe* by Kate Pullinger (2018), for example, is a story about a young woman who talks to ghosts. The opening line of the story is, 'Pick up your phone, I'm ready'. Images of the reader's location, captured through the phone's camera, appear during the story, as do the reader's geographical location and local weather conditions. The reader also has to move the phone in order to make (hidden) text appear. The overall effect is disorientating, as the reader's own world slowly creeps into the fictional story at the same time as they are interacting with the phone.

From a story like *Breathe*, with its layering of fictional graphics and real-world video and data capture, it isn't a big step to the use of either AR or VR in storytelling. Whereas in *Breathe* the reader is still very much *within* a textual story in which the real world plays a walk-on part, with AR or VR the situation is reversed. In AR, computer-generated enhancements are layered on top of the real world through an electronic device, such as a smartphone or iPad. A growing range of AR developer platforms (including Apple's ARKit) are slowly simplifying both the cost and the construction of such experiences. A good example is *Pokémon Go*, an app released in July 2016, which overlays gameplay on top of real world locations. AR books are slowly gaining traction too, particularly within the children's market. Here, an app is used alongside a traditionally printed book. The reader points their device at certain passages or illustrations in the book and their device shows additional content such as a video or an interactive animated sequence. Not surprisingly, AR has also become popular within the heritage and museum sector too, where additional information or stories can be overlaid on real world objects or locations. At the Jinsha Site Museum in Chengdu, China, visitors are invited to use an app to explore the stories behind various artefacts and relics of the lost civilisation known as Shu; while *England's Historic Cities* app uses AR technology to tell the stories of various heritage sites, including York, Oxford and Bath.

Whereas AR layers digital enhancements over the real world, VR immerses the user within a totally simulated world. According to Pimentel and Teixeira (1993), it is this immersivity, alongside interactivity, that defines the VR experience. This is most commonly achieved through special headsets, such as the HTC Vive or Oculus Rift, that could also include eye tracking software and data gloves. The overall effect is to give the user the sense that they are experiencing the simulated reality first hand. Although the equipment can be expensive, it certainly doesn't have to be. Google Cardboard relies on a simple cardboard viewer into which a smartphone is inserted. The experience itself is delivered by one of the many VR apps downloaded from Google Play. Yet for now, at least, the best experiences are to be had at the most expensive end of the spectrum. *Bloodless* (Kim 2017), a ten-minute VR film by the South Korean documentary filmmaker, Gina Kim, follows the last moments of a sex worker who was murdered by a US soldier at the Dongducheon Camptown in South Korea in 1992. *Sea Prayer* (Hosseini 2017) is a VR animated story by the award-winning novelist Khaled Hosseini, commemorating the death of Alan Kurdi, the three-year-old Syrian boy who drowned while attempting to reach Greece in 2015. The story takes the form of an imagined letter, written by a Syrian father to his son lying asleep in his lap, on the eve of their sea crossing to Europe.

A similar level of experimentation is taking place in contemporary drama. The Australian Dance Theatre, in cooperation with technology companies *Sandpit* and *Jumpgate*, have created a VR version of their dance and music performance, *The Beginning of Nature*. The VR experience is delivered through an app that works with Google Cardboard, allowing users to view a two-minute 360 degree video of the choreography filmed in a variety of rural locations around Adelaide, Australia. More ambitiously is a project such as *Empire Soldiers VR* by the theatre company Metro Boulot Dodo, UK. The company calls *Empire Soldiers VR* an 'experience installation' because it blends both virtual and real-life performance elements. The project seeks to raise awareness of the history of Caribbean soldiers during the First World War. As well as experiencing real-life dancing, the audience undertakes a virtual journey accompanied by a Caribbean soldier returning from the front line in which they hear stories of the battlefield. The journey continues up to the present day, allowing the audience to reflect on contemporary migration stories.

For Gabriella Giannachi and Nick Kaye, this growing interdependence between contemporary performance and what they call 'new media forms', brings into question established notions of presence within performance theory. They note how these new technologies facilitate the contest between 'absence' and 'presence' within a performance: 'between a property or quality of "a" presence "belonging" or manifest *in* the body, as a "real" act, or as an immanent property of an object or thing, and the "illusion" of presence effected in a staging of forms and modes of representation' (2017, 241). This tension between absence and presence, between what is understood as real and not real, corporeal and non-corporeal, will reach out across this book. Tellingly, Manovich uses the term 'software performances' (2013, 33) to describe the way electronic systems themselves now work: their output, such as web pages and apps, are inherently dynamic, contingent and generated in real time. There's a sense then that any concept of performance and presence should be seen as something akin to an ecosystem, embracing not only the performers, the audience and the media, but also the underlying (performative) characteristics of the very software itself.

The use of VR and AR as a means of telling stories is still in its early days. Fundamentally, there still remains a tension between immersion, interactivity and narrativity that we've already met, the balance between the real-life freedom proffered by immersion and interactivity, and the need for narrative structure and design. As Ryan notes, it's a combination 'whose formula still eludes us' (2015, 259).

This pervasive digital presence is all part of our postdigital age. The computer has become, what Manovich calls, a metamedium, where different programmes share common software techniques and design (2013, 123). A key part of this phenomenon is the rise of 'media hybridization' in which existing media technology become 'building blocks for many new media combinations' (2013, 163). A website, for example, may combine photographs, animation, live social media feeds and interactive elements. This is not simply 'multimedia' by another name. Manovich describes how, although multimedia blended different media together, each still opened using their own specific software (2013, 167). The combination of hybridized media is far-more fundamental: 'In hybrid media the languages of previously distinct media come together. They exchange properties, create new structures, and interact on the deepest levels' (2013, 169). The postdigital pushes this even further. Here, it is not just digital hybridity that comes into play but rather the new found entanglement between the digital and the non-digital, in other words the transmediatised world in which we live. As this book argues, these developments, from multimedia capability to media hybridisation, from media hybridisation to postdigital transmeditisation, has brought with it profound and significant creativity opportunity. In this sense, the book takes up the challenge thrown down by N. Katherine Hayles in her demand for a new kind of approach to the study of digital media, one 'that can locate digital work within print traditions, and print traditions within digital media, without obscuring or failing to account for the differences between them' (2012, 7).

Yet, if this is indeed a postdigital age, then, as we've seen, it is also the age of the story. In part, this is being driven by the digital technology explored in this chapter and elsewhere in this book. The proliferation of the smartphone, alongside other handheld devices such as iPads and ereaders, have revolutionised how we access and read stories. Yet, at the same time, the interdependency and hybridisation of software, has provided a vast array of new creative potential, from blogs and wikis, proprietary storifying platforms such as Twine through to the timelines and threads of Facebook and Twitter. Never before have there been so many opportunities for writers to create and publish their work electronically (Hammond 2016, Chapter 6). For the first time in history, individuals and communities have been given the technical affordances to create and share their own stories across the globe.

The impact of this is difficult to overestimate. There are downsides of course – the proliferation of fake news is a consequence of many things, including how corporate platforms such as Facebook, Instagram and Twitter ghettoise news streams through homophilous sorting (D'Ancona 2017, 50).

There can be no doubt that the deliberate spreading of false or malicious lies is an attempt to weaponise social media in the battle for social and political power (*The Economist* 2017). Yet the negative publicity surrounding post-truth, Oxford Dictionaries Word of the Year for 2016, should not detract from the enormous benefit that individual and community storytelling can bring. One of the underlying causes of the post-truth/fake news syndrome is surely that, within a globalised world, the role of the individual, and the communities in which she or he lives, are easily forgotten, subsumed within the vast, macro world view of economic and social policy. If stories have a role here it's to remind ourselves how the world is actually experienced: individually, through the local and the intimate. As I've already shown, from issues such as global warming, through to urbanisation, aging and social resilience, the power of stories is slowly being recognised. A good example of this comes out of the UK-based Connected Communities Programme launched by Research Councils UK and the Arts and Humanities Research Council. The £30 million programme has the aim of funding projects, partnerships and networks that explore the historic and contemporary condition of 'community'. As Keri Facer and Bryony Enwright make clear, a major influence on the programme has been, and continues to be, what they call the rise of the 'participatory turn' in knowledge creation, 'in which users/publics/patients/audiences/communities are invited to take on more active roles in shaping the knowledge, policies and practices of the world around them' (2016, 144). As they go on to note, storifying is central to this process: 'how we produce knowledge, scholarship and ideas about reality matters for the stories we are able to tell about ourselves and our society, and for how we frame our response to the changing realities of the contemporary world (2016, 11). Further afield, the Sidney Myer Fund and The Myer Foundation in Australia has a similar focus while, in the United States, the National Endowment for the Arts has grant programmes aimed at the integration of arts, culture and community. The 'Our Town' grant program, for example, specifically supports creative placemaking projects, bringing together artists, arts organisations and community development practitioners to help transform local lives.

Computationalism and storytelling

Yet it would be wrong to deny that there isn't a tension here. If normalisation has brought with it all sorts of technological opportunities for storytelling, there's also a danger that with it comes a blindness to the inherent power structures enacted through digital systems (Tabbi 2018; Heckman and O'Sullivan 2018). For a critic such as David Golumbia, this sort of deep socio-cultural enmeshing, what he calls 'computationalism', brings with it fundamental dangers and risks.

One powerful tradition within digital storytelling was (and remains) the creation of texts that specifically illuminate and reveal the iniquitous effects of 'computationalism' or what John Cayley calls 'Big Software', in other words the deep and pervasive influence of computational power and the global companies that lie behind it. In *A Hacker Manifesto* (2004) McKensie Wark transposes these concepts of technological resistance onto a Marxist framework, in which the classic nineteenth and twentieth-century class conflict between the proletariat and the bourgeoisie is reconfigured into that between the 'vectoralists', those who control flows of information, and the hacker class, who enable innovation and open up new information flows.

Golumbia takes a different position, cautioning against using computer systems at all in the wider critique of corporate and institutionalised power structures. Although he notes the raise of 'new media art, hackers and open source software' (2009, 4), and recognises that, at the level of the individual they can be empowering, Golumbia states that it is a mistake to see them as offering any real impact on, what he describes as, the inherent authoritarianism of computerisation (2009, 220).

It's an extreme position. Taken literally it implies that nothing less than a rollback of computerisation across the planet would offer an escape from its more iniquitous effects (a utopian rally cry that Golumbia makes in his book). In other words, computers and computer systems are inherently bad and the only way of escaping their influence, and the power structures they support, is by switching them off. This is clearly an important argument and Golumbia isn't the only person to have made it (for example, see Srnicek 2017; Mosco 2005; Schiller 2000). No book that seeks to understand the relationship between creativity and digital systems can afford to ignore such fundamental concerns. At least two questions are useful here though: first, is there truth to the claim that computer systems are indeed in thrall to the sort of authoritarian hegemony described by critics such as Golumbia and Srnicek; and second, even if the first question is true, where does that leave the role of creativity, specifically that enacted through those very computer-based systems that are seen to be instrumental in authoritarian systems of control?

As we have seen, many early practitioners of digital storytelling, those working in the 1990s for example, saw their practice as, to use Ciccoricco's term, a 'kind of art of resistance' where electronic works were an explicit 'mode of political or ideological intervention' (2018, 151). In his 2004 book Alexander Galloway called hackers 'freedom fighters' (2004, 152), a metaphor that extended beyond those writing code, to those producing digital art capable of turning computerised systems against themselves.

Yet, now, as the 1990s and those early years of the millennium recede into the middle distance, critics such as Tabbi are becoming less sanguine about this radical tradition. In a world where computerisation is fundamentally normalised, any understanding of the digital as subversive and radical becomes redundant; for Tabbi, this postdigital world 'is no place for avant-gardes' (2018, 7). Instead, he argues that if electronic literature is now mainstream then it is from here (rather than from any avant-garde periphery) that writers should evolve a 'post-digital poetics' capable of defamiliarising our digitised condition. And in so doing, making what was invisible visible again.

Although writers such as Golumbia would argue this is impossible *tout court*, this book takes up Tabbi's challenge by seeking to explore both the practical and the philosophical limits of a postdigital poetics. Galloway, in his more recent study, analyses the digital condition through the work of the French philosopher, François Laruelle. Galloway notes that, for Laruelle, an event (defined as a relation between two moments in time) has two vectors: a movement of *reality* and a movement of *freedom*, each working in opposite directions (2014, 80). For Laruelle, the movement of freedom seeks an 'elevated artifice' (what he calls fictions, artifices or performations) using data 'as they are assembled within relational and decisional events' (2014, 80). As Galloway explains, 'digital freedom is thus a question of being "free from" the autonomy of data. Counterintuitively, then, the movement of freedom is driven not by liberation but by increased imbrication with the sociopolitical sphere' (2014, 81). In this way, the sociopolitical sphere (to use Galloway's term) is forced to engage with (to be encoded within) the event, and in that way 'move closer to freedom' (2014, 81). Freedom can only come from this entrenched position.

Yet there's also other, equally important, questions that need addressing here. For example, behind Tabbi's argument lies the unspoken assumption that the overt role of any digital literature should be to critique and engage ideologically with systems of power. I would argue that this sort of argument exhibits a dangerous reductionism. Perhaps at its heart lies an assumption about intent – that a radical text is simply determined by the intention of the writer. Surely the reality is far-more complex. What a reader or an audience draws from a text, whether digital or not, is ultimately beyond the control of the writer. Did an author such as Jane Austen (1775–1817) consciously set out to challenge both the form and language of the novel, not to mention gender-based relations? Did Art Spiegelman in *Maus* (1991) sit down to write a story that would both transform people's opinion of graphic novels, as well as their understanding of history and personal testimony? I would suggest that in each of these cases the answer is no. The primary drive for both Jane Austen and Art Spiegelman was to tell their story in the way that felt right to them. A radical text then, one that challenges established ideological conditions, is one that is chosen to be so by its readership, rather than its author. As writers and artists, we should be careful of falling into the trap of thinking that it is us who ultimately decide the role our texts should and will play.

There are two other points that are also relevant here. The first is not to forget the skills and knowledge needed to actually create and distribute texts in our postdigital age. Even in the traditional world of publishing, almost the entire process, apart from the actual reading of the hardcopy book, is done electronically. Unlike twenty years ago, a would-be author today needs at the very least a good understanding of word processing. And this skills bar is raised even higher once we start to discuss digital texts, in other words stories made explicitly for a digital interface. As we'll see further on in this book, digital stories invite authors to experiment with coding (simple or complex html for example, or platform-specific commands such as those used by Twine or an app development interface), to explore the complexities of human-computer interaction and the limits and possibilities of digitised narrative. These are not inconsequential skills and knowledge; if there is a radical intent in the creation of a digital text then surely it, at least in part, lies here, in giving ordinary women and men the ability to move from the passive use of computerisation (such as word processing), to the active state of playful experimentation.

The second point I'd like to raise relates to what I've already said in terms of the wider role and function of creativity, particularly storytelling. As we've seen with the rise of the 'participatory turn' (Facer and Enwright 2016), storytelling, or storifying, is now recognised as a fundamental, and transformative, human activity. In other words, stories are a big part of how we make sense of the world (Barrett and Bolt 2010). This knowledge production happens at many levels, from the individual and community, through to the city, nation and the global (Hawkins 2017). In fact, this community-focussed approach forms a completely separate tradition within the history of digital storytelling (see Lambert 2013; Dunford and Jenkins 2017). The *Center for Digital Storytelling*, based at the University of California, Berkeley, was established in 1993, and their goal, then and now, is very much about exploring the way digital storytelling can be used to enhance the power of the personal, community-situated, voice as an instrument of change. As I've already begun to argue, software that enables ordinary people to construct and share their stories can be transformative. The fact that such systems are part of Golumbia's 'computationalism' should not distract from the benefit that such storytelling accords us, especially in view of the many challenges ahead. Democracy is on the decrease around the world with more than two dozen countries having reverted to some form of authoritarianism this century (Mair 2013; *Freedom in the World* 2018). In many of these authoritarian states, in countries such as Russia, China, Turkey and Hungary, it is the control and limitation of digital platforms which forms a cornerstone of their autocracy (*Freedom on the Net* 2017*: Manipulating Social Media to Undermine Democracy* 2017). It is these platforms' ability to circumvent official power structures, to empower and liberate in multiple ways, that makes them such a threat to authoritarian rule. So, yes, critics such as Golumbia are certainly right in highlighting the dangers of globalised computationalism; yet, that is only half the story. It is that very computationalism, embracing countless individuals across a borderless global network that makes it so dangerous (and ultimately anathema) to authoritarian rule. Our 'post-truth/fake news' condition is a product of repeated and ongoing state-sponsored attempts to turn digital platforms against this liberal, democratic tradition (D'Ancona 2017; Mair 2013).

The challenges ahead are not just political ones, of course. We live in a time of unprecedented social, ecological and climatic disruption. Although not formally recognised, 'Anthropocene' is increasingly accepted as describing a new era in human history, in which landscape degradation, urbanisation, species extinction and resource extraction are endemic (Davies 2016). Quite naturally, scientists have played starring roles in understanding such a complex phenomenon. Yet it would be wrong to think that the arts and humanities do not also have a crucial role to play here as well. After all, the effects of the Anthropocene are not limited to the landscape and climate but ripple across all aspects of human life. As Clive Hamilton, Christophe Bonneuil and François Gemenne state, 'reinventing a life of dignity for all humans in a finite and disrupted Earth has become the master issue of our time' (2015, 5). This book argues that it is here, in this reformulation of a 'life of dignity' in the face of mounting social, cultural, economic and political pressure, that a postdigital poetics can play an important role. In his book, *Molecular Red: Theory for the Anthropocene* (2015), McKensie Wark outlines how this might be done. Critical here is what he labels 'low theory', in other words how things such as concepts and stories are actually embedded in practice: 'low theory is interstitial, its labour communicative rather than controlling' (2015, 218). I would argue that constructing and maintaining such low theory solutions to twenty-first century living will in part be achieved through emancipatory and revelatory acts of (postdigital) creativity. As Lambert notes, 'through digital storytelling, we can all become storytellers again' (2013, 5).

Themes and structure

Having summarised some of the major issues that are pertinent to what I'm going to say, I'd now like to briefly set out how the text is arranged. It is worth repeating here that the focus for the book is storytelling within the arts and humanities. As such, it also takes an international perspective on both the application of practice-based research and storytelling as a nexus of new and emerging practice. By storytelling I refer to prose-based work, both fiction and non-fiction. Although poetry has a strong tradition within digital storytelling, it is not an explicit focus of my study, although much that I have to say remains independent of any literary form and genre.

The heart of the book remains a critical analysis of current trends and features in the use of, what I have termed, postdigital storytelling. But before I do that, it's important to provide some necessary context that will then inform any analytical framework to be used across the rest of the chapters. This is the intention behind the opening section, 'Pasts and Presents'. It consists of four chapters, the first two of which (Chapters 2 and 3) place discussions of contemporary storytelling within a wider, and deeper, debate about creativity itself. In Chapter 2, rather than understanding creativity as a monolithic given, I show how the ontology that underpins artistic representation has gone through at least three paradigmatic shifts, what Richard Kearney identifies as mirror, lamp and labyrinth (1988). I argue that with the ending of postmodernity, we have now entered a new, fourth era of creative modality, building on Martin Heidegger's (1889–1976) concept of 'worlding the world'. Key here are creative interventions that are embodied, ethically-charged and affective in their engagement with the world. Chapter 2 uses this theoretical framework to advance a reconfigured understanding of creativity within the arts and humanities; it also outlines how an affective and embodied understanding of creativity leads us towards interdisciplinary and transdisciplinary approaches to practice-based research. Chapter 3 explores how the concept of the postdigital offers new and exciting ways of rethinking our post-postmodern condition. It sets out a new and modified understanding of postdigitality, before exploring its impact on established conceptual terms such as digital storytelling and electronic literature. As well as positing a revised four-layer model of postdigital storytelling, building on Katherine N. Hayles' print/code dichotomy (2004), the chapter examines the role and function of storytelling as a form of research, establishing an enabling relationship between practice, praxis and research.

Chapters 4 and 5, 'Hypertextual Adventures' and 'Hypertextual Storytelling Today', analyse the rise and subsequent evolution of hypertext fiction, from classic works such as Michael Joyce's *afternoon: a story* (1987) through to the development of Twine and the hypertextual storytelling of research agencies such as Forensic Architecture. In this way they provide the second axis to my analytical framework, namely, longitudinal and historical, charting the development of hypertext fiction from postmodern disorientation to the affective earnestness of metamodernism. As I've already indicated, the late twentieth century was a period of profound technological change, witnessing both the rise of the personal computer and the smartphone; hypertext fiction was just one means by which artists were able to respond to these transformative innovations. This was nothing new, of course. The nineteenth and early twentieth centuries had already shown that the impact of any technological upheaval could be profound, prefiguring deep and sustained cultural and social transformation. The 'crisis of the senses' (Danius 2002, 3) that underpinned modernist literary experimentation in the first half of the twentieth century, for example, was in part a response to the disruptive effects of the telephone, radio and cinema. By the late 1980s, digital innovation such as hypertext had vastly increased the creative options for any artist. Yet the exploration of hypertextuality as a creative affordance was not limited to the web. Chapter 4 argues that Mark. Z. Danielewski's *House of Leaves* (2000) was one of the first printed novels to explore a new kind of transmedial hybridity, positioning the physical book within a network of paratextual digital material. At the same time, hypertext works such as William Gillespie, Scott Rettberg and Dirk Stratton's *The Unknown* (1998) and *The Doll Games* (2001) by Shelley and Pamela Jackson, embraced new fusions of participatory and performative storytelling. Chapter 5 shows how, by the second decade of the new millennium, hypertextuality itself had become a metaphor for new, hybridic forms of transmedial interconnectivity. From J. K. Rowling's *Pottermore* website to Richard House's use of web-based video in his novel, *The Kills*

(2013), digital storytelling had become increasingly postdigital in form and structure. In this condition of normalised hybridity, any artistic work that remains disengaged from the digital domain does so consciously and tactically. Matthew McIntosh's *theMystery.doc* (2017), an analogue 'hypertextual' novel, is a case in point. Yet behind all these works is evidence of a new ontological imperative, a turning away from postmodern scepticism towards an ethically-informed 'worlding' (Trend 2016).

The second section of the book, 'Into Infinity – Towards a Postdigital Poetics', focuses on the tactics, strategies and conventions of contemporary storytelling. Chapter 6, 'Spatiality and Text: Locative Mobile Storytelling', examines the very recent phenomenon of mobile storytelling, in other words storytelling in which the reader's physical journey becomes a key aspect of a (performative) story. The chapter argues that this represents a stepwise change in how writers think about narrative. Key to this transformation is the creation of what I call *embodied space*. Embodied space is a hybridic form of narrative space that foregrounds the situated embodiedness, the essential postdigital entanglement, connecting all our lives. The chapter argues that it is through embodied space that locative mobile storytelling engages with metamodernist sensibility, particularly in regard to autofiction.

Chapter 7, 'Collaborative Tales', explores two ways that authors are engaging with participatory and collaborative approaches. The first is an examination of the Centre for Digital Storytelling, with its emphasis on co-creation and empathy through such techniques as the story circle within a seven-step creative process (Lambert 2013). Critical here is the notion of digital storytelling as a 'learning modality' (Lambert 2013, 14). The chapter then moves on to look at the ways in which social media can be used within digital arts and humanities' research. Although recognising challenges and complexities in the way social media has impacted on society, the chapter argues that there still remains significant opportunity for researchers and authors alike to harness the enormous collaborative potential of such platforms.

The final chapter of the book, 'How Soon Is Now', looks both backwards in terms of what has been said about postdigital storytelling in the previous chapters, but also forwards in terms of cutting-edge developments such as AI and the sort of machine-to-machine interactions captured by Trevor Paglen's *How to See Like a Machine* project (2017). Indeed, artists such as Paglen remind us that we are fast approaching a time when the majority of machine activity will not involve any direct human input at all. If digital storytelling increasingly consists of these kinds of cybernetic stories that machines tell each other, postdigital storytelling has become the imperative by which we can champion the 'being-in-the-world' of the human condition. The chapter argues that transdisciplinarity is critical here, providing an approach by which the essentiality of this human experience is foregrounded. As the chapter notes, postdigital storytelling and transdisciplinarity are irrevocably intertwined, each the child of a new ontological paradigm, a paradigm I've termed metamodernism.

Storytelling has much to offer the research of global imperatives such as resilience and empathy (Bazalgette 2017), yet without the means to reach across traditional academic boundaries, we risk being overwhelmed by the sheer scale of what faces us.

Summary

This chapter has introduced, what I consider to be, key areas for debate and analysis in regards to postdigital storytelling. Perhaps the most important aspect of this has been the introduction of postdigital to describe the critical context through which this analysis will be framed. Most fundamentally, I have outlined how I will expand and deepen the concept of postdigitality, elevating it from a straightforward description of technological hybridity to a more complex socio-technological phenomenon that is itself a primary mode of a new and emergent cultural paradigm.

I've highlighted the degree to which the speed and depth of technological change since the turn of the century provides the central pillar on which this phenomenon rests. Yet equally importantly, I've also stressed the concomitant rise in the perceived value and utility of storytelling *per se* within the arts and humanities. Any understanding of a postdigital poetics as an emergent, transformative praxis, needs to at least recognise these two interrelated phenomena.

The chapter has indicated how I'm going to be using my critical engagement with postdigital storytelling to discuss the broader issues surrounding practice-based research, a methodological approach that is still significantly under investigated. This emergent nexus between creativity (poetics) and practice-based research (praxis) forms the methodological focus of this book in which both poetics and praxis play interdependent explanatory roles.

Finally I began to unpick the close relationship that is often presumed to exist between digital storytelling and the avant-garde. I argued that this was an outmoded view of storytelling *per se*. If an imperative exists for postdigital storytelling it is better to look at how such narrative approaches help address wider issues within our society. I introduced the concept that postdigital storytelling can be seen as a critical focus through which to advance a more transformative framework for transdisciplinary research across the arts and humanities and beyond.

Whether or not the Anthropocene is formally recognised as a geological epoch, we live in a world of increasing political, social, economic and environmental distress. Exploring the complex ways in which knowledge is created is of vital importance, and at the heart of this lies the stories we tell ourselves. As John Hartley says, '[f]ictions not only bond groups; they "imagine" the most compelling realities we live by' (2017, 218). Understanding how postdigital storytelling may help conjure up and instantiate such 'compelling realities' is, at least in part, a focus of the following chapters.

Note

1 Some authors use the hyphenated *post-digital*. I prefer, and use, the unhyphenated *postdigital* throughout this book.

Works cited

van den Akker, R. Gibbons, A. and Vermeulen, T., eds. *Metamodernism: Historicity, Affect, and Depth After Postmodernism*. London: Rowman & Littlefield International, 2017.

Alexenberg, M. *The Future of Art in a Postdigital Age: From Hellenistic to Hebraic Consciousness*. Bristol: Intellect, 2011.

Anderson, S. *Technologies of History: Visual Media and the Eccentricity of the Past*. Boston: Dartmouth College Press, 2011.

Barrett, E. 'Introduction'. *Practice as Research: Approaches to Creative Arts Enquiry*. Eds, Barrett, E. and Bolt, B. London: I.B. Tauris, 2010, 1–14.

Barrett, E. and Bolt, B., eds. *Practice as Research: Approaches to Creative Arts Enquiry*. London: I.B. Tauris, 2010.

Baum, L. Frank. *The Patchwork Girl of Oz*. Chicago: Reilly & Britton, 1913.

Bazalgette, P. *The Empathy Instinct: How to Create a More Civil Society*. London: John Murray, 2017.

Bell, A *The Possible Worlds of Hypertext Fiction*. Basingstoke: Palgrave Macmillan, 2010.

Bernbeck, R. 'Narrations in Archaeology: From Systems to People'. *Subjects and Narratives in Archaeology*. Eds, van Dyke, R. M. and Bernbeck, R. Boulder, Colorado: University Press of Colorado, 2015, 277–286.

Berry, D. M. and Dieter, M. 'Thinking Postdigital Aesthetics: Art, Computation and Design'. *Postdigital Aesthetics: Art, Computation And Design*. Eds, Berry, D. M. and Dieter, M. London: Palgrave Macmillan, 2015, 1–11.

Bernstein, J. H. 'Transdisciplinarity: A Review of its Origins, Development, and Current Issues'. *Journal of Research Practice*, 11 (1), 2015. Web. 15 February 2018. http://jrp.icaap.org/index.php/jrp/article/view/510/412.

Boyd, B. *On the Origin of Stories: Evolution, Cognition, and Fiction*. Cambridge, Massachusetts: The Belknap Press of Harvard University Press, 2009.

Cameron, E. 'New Geographies of Story and Storytelling'. *Progress in Human Geography*, 36 (5), 2012: 573–592.

Cassidy, L. 'Salford 7/District Six. The Use of Participatory Mapping and Material Artefacts in Cultural Memory Projects'. *Mapping Cultures: Place, Practice, Performance*. Ed, Roberts, L. London: Palgrave Macmillan, 2015, 181–198.

Cayley, J. 'The Advent of Aurature and the End of (Electronic) Literature'. *The Bloomsbury Handbook of Electronic Literature*. Ed, Tabbi, J. London: Bloomsbury, 2018, 73–91.

Ciccoricco, D. 'Digital Fiction: Networked Narratives'. *Routledge Companion to Experimental Literature*. Eds, Bray, J., Gibbons, A. and McHale, B. London: Routledge, 2012, 469–482.

Ciccoricco, D. 'Rebooting Cognition in Electronic Literature'. *The Bloomsbury Handbook of Electronic Literature*. Ed, Tabbi, J. London: Bloomsbury, 2018, 151–164.

Cooper, D. Donaldson, C. and Murrieta-Flores, P., eds. *Literary Mapping in the Digital Age*. London: Routledge, 2017.

Cramer, F. 'Afterword'. *Post-Digital Print: The Mutation of Publishing Since 1894*. Ludovico, A. Eindhoven. The Netherlands: Onomatopee, 2012, 162–166.

Cramer, F. 'What is "Post-Digital"?'. *Postdigital Aesthetics: Art, Computation and Design*. Eds, Berry, D. M. and Dieter, M. London: Palgrave Macmillan, 2015, 12–26.

D'Ancona, M. *Post-Truth: The New War on Truth and How to Fight Back*. London: Ebury Press, 2017.

Danielewski, M. Z. *House of Leaves*. London: Transworld Publishers, 2000.

Danius, S. *The Senses of Modernism: Technology, Perception and Aesthetics*. Ithaca, New York: Cornell University Press, 2002.

Davies, J. *The Birth of the Anthropocene*. Oakland, California: University of California Press, 2016.

Dunford, M. and Jenkins, T., eds. *Digital Storytelling: Form and Content*. London: Palgrave Macmillan, 2017.

The Economist. 'Once Considered a Boon to Democracy, Social Media have Started to Look like its Nemesis'. *The Economist*, 4 November (2017). Web. 18 January 2018. www.economist.com/news/briefing/21730870-economy-based-attention-easily-gamed-once-considered-boon-democracy-social-media.

The Economist Intelligence Unit. *Digital Upheaval: How Asia-Pacific is Leading the Way in Emerging Media Consumption Trends*. London: The Economist Intelligence Unit, 2017.

Editions at Play. Web. 5 February 2018. https://editionsatplay.withgoogle.com/#/about.

Ensslin, A. *Literary Gaming*. Cambridge, Massachusetts: Massachusetts Institute of Technology, 2014.

Facer, K. and Enwright, B. *Creating Living Knowledge: The Connected Communities Programme, Community-University Relationships and the Participatory Turn in the Production of Knowledge*. Bristol: University of Bristol and AHRC Connected Communities Programme, 2016.

Facer, K. and Pahl, K. 'Introduction'. *Valuing Interdisciplinary Collaborative Research: Beyond Impact*. Eds, Facer, K. and Pahl, K. Bristol: Polity Press, 2017. 1–21.

Felt, U. 'Making Knowledge, People, and Societies'. *The Handbook of Science and Technology Studies*. Eds, Felt, U., Fouché, R., Miller, C. A. and Smith-Doerr, L. Cambridge, Massachusetts: Massachusetts Institute of Technology, 2017, 253–257.

Foucault, M. 'Of Other Spaces'. *Diacritics*, 16 (1) 1986: 22–27.

Freedom on the Net 2017: Manipulating Social Media to Undermine Democracy. Washington D. C.: Freedom House, 2017.

Freedom in the World. Washington D. C.: Freedom House, 2018.

Gabbay, J. and Le May, A. *Practice-based Evidence for Healthcare: Clinical Mindlines*. London: Routledge, 2011.

Galloway, A. R. *Laruelle: Against the Digital*. Minneapolis: University of Minneapolis Press, 2014.

Galloway, A. R. *Protocol: How Control Exists after Decentralization*. Cambridge, MA: The MIT Press, 2004.

Gauntlett, D. *Making Media Studies: The Creativity Turn in Media and Communications Studies*. New York: Peter Lang Publishing, 2015.

Giannachi, G. and Kaye, N. *Performing Presence: Between the Live and the Simulated*. Manchester: Manchester University Press, 2017.

Giannachi, G., Kaye, N. and Shanks, M. 'Introduction: Archaeologies of Presence'. *Archaeologies of Presence: Art, Performance and the Persistence of Being*. Eds, Giannachi, G., Kaye, N. and Shanks, M. London: Routledge, 2012. 1–25.

Gillespie, W., Rettberg, S. Stratton, D. and Marquardt, F. *The Unknown*. 1998. Web. 30 April 2018. http://unknownhypertext.com.

Golumbia, D. *The Cultural Logic of Computation*. Cambridge, Mass.: Harvard University Press, 2009.

Graham, C. 'Crowd Turns its Back on Hillary Clinton as Photo Captures the Age of the Selfie'. *The Daily Telegraph*, 26 September (2016). Web. 26 January 2019. www.telegraph.co.uk/news/2016/09/25/crowd-turns-its-back-on-hillary-clinton-as-photo-captures-the-ag.

Hamilton, C., Bonneuil, C. and Gemenne, F. 'Thinking the Anthropocene'. *The Anthropocene and the Global Environmental Crisis*. Eds, Hamilton, C., Bonneuil, C. and Gemenne, F. London: Routledge, 2015, 1–13.

Hammond, A. *Literature in the Digital Age: An Introduction*. Cambridge: Cambridge University Press, 2016.

Haraway, D. J. *Staying with the Trouble: Making Kin in the Chthulucene*. Durham: Duke University Press, 2016.

Harris, A. *The Creative Turn: Toward a New Aesthetic Imaginary*. Rotterdam: Sense Publishers, 2014.

Hartley, J. 'Smiling or Smiting? – Selves, States and Stories in the Constitution of Polities'. *Digital Storytelling: Form and Content*. Eds, Dunford, M. and Jenkins, T. London: Palgrave Macmillan, 2017. 203–227.

Hawkins, H. *Creativity*. London: Routledge, 2017.

Hawlin, T. 'No One Gets Out Alive: An Interview with Joanna Walsh'. *Review 31*. 2017. Web. 1 February 2018. http://review31.co.uk/interview/view/19/no-one-gets-out-alive-an-interview-with-joanna-walsh.

Hayles, N. K. *How We Think: Digital Media and Contemporary Technogenesis*. Chicago: University of Chicago Press, 2012.

Hayles, N. K. 'Print is Flat, Code is Deep: The Importance of Media-Specific Analysis'. *Poetics Today*, 25 (1), 2004: 67–90.

Heckman, D. and O'Sullivan, J. '"your visit will leave a permanent mark": Poetics in the Post-Digital Economy'. *The Bloomsbury Handbook of Electronic Literature*. Ed, Tabbi, J. London: Bloomsbury, 2018, 95–112.

Horst, H. A. and Miller, D. 'The Digital and the Human: A Prospectus for Digital Anthropology'. *Digital Anthropology*. Eds, Horst, H. A. and Miller, D. London: Berg, 2012, 3–36.

Hosseini, K. *Sea Prayer*. London: Guardian VR, 2017.

House, R. *The Kills*. London: Picador, 2013.

Jackson, S. *Patchwork Girl; or a Modern Monster by Mary/Shelley and Herself*. Watertown, MA: Eastgate Systems, 1995.

Jackson, S. and Jackson, P. *The Doll Games*, 2001. Web. 28 May 2018. www.ineradicablestain.com/dollgames.

Johnson, B. S. *The Unfortunates*. London: Panther Books, 1969.

Joyce, M. *afternoon: a story*. Watertown, MA: Eastgate Systems, 1987.

Kearney, R. *Poetics of Imagining: Modern to Post-Modern*. Edinburgh: Edinburgh University Press, 1998.

Kim, G. *Bloodless*. [Film]. South Korea: Crayon Film Production, 2017.

Koehler, A. *Composition, Creative Writing Studies, and the Digital Humanities*. London: Bloomsbury, 2017.

Lambert, J. *Digital Storytelling: Capturing Lives, Creating Community*. London: Routledge, 2013.

Lefebvre, H. *The Production of Space*. Oxford: Blackwells, 1991.
Losse, K. 'The Return of the Selfie'. *The New Yorker*. 5 June 2013. Web. 17 January 2018. www.newyorker.com/tech/elements/the-return-of-the-selfie.
Lucas, G. 'Modern Disturbances: On the Ambiguities of Archaeology'. *Modernism/Modernity*, 11 (1), 2004: 109–120.
Lupton, D. *Digital Sociology*. London: Routledge, 2014.
McHale, B. *Postmodernist Fiction*. London: Routledge, 1999.
McIntosh, M. *theMystery.doc*. London: Grove Press, 2017.
McMullan, T. 'Word Games: How Technology can Enrich Storytelling'. *Times Literary Supplement*. 26 May 2017: 22.
McQuire, S. *The Media City: Media, Architecture and Urban Space*, London: Sage, 2008.
Mair, P. *Ruling The Void: The Hollowing of Western Democracy*. London: Verso, 2013.
Malpas, J. E. *Place and Experience: A Philosophical Topography*. Cambridge: Cambridge University Press, 1999.
Manovich, L. *The Language of New Media*. Cambridge: MIT Press, 2001.
Manovich, L. *Software Takes Command*. London: Bloomsbury, 2013.
Mosco, V. *The Digital Sublime: Myth, Power, and Cyberspace*. Cambridge, MA: The MIT Press, 2005.
Naughton, J. *From Gutenberg to Zuckerberg; What you Really Need to Know about the Internet*. London: Quercus, 2012.
Ofcom. *The Communications Market Report*. London: Ofcom, 2017.
Perloff, M. *Unoriginal Genius: Poetry by Other Means in the New Century*. Chicago: University of Chicago Press, 2010.
Pimentel, K. and Teixeira, K. *Virtual Reality: Through the Looking Glass*. New York: Intel/Windcrest McGraw Hill, 1993.
Pullinger, K. *Breathe*. Editions at Play, 2018. Web. 5 February 2018. https://editionsatplay.withgoogle.com/#/detail/free-breathe.
Punday, D. 'Narrativity'. *The Bloomsbury Handbook of Electronic Literature*. Ed, Tabbi, J. London: Bloomsbury, 2018, 133–149.
Rajan, R. *The Third Pillar: The Revival of Community in a Polarised World*. London: HarperCollins, 2019.
Roberts, L. 'Mapping Cultures: A Spatial Anthropology.' *Mapping Cultures: Place, Practice, Performance*. Ed, Roberts, L. London: Palgrave Macmillan, 2015. 1–25.
Ryan, M. *Cyberspace Textuality: Computer Technology and Literary Theory*. Bloomington: Indiana University Press, 1999.
Ryan, M. *Narrative as Virtual Reality 2: Revisiting Immersion and Interactivity in Literature and Electronic Media*. Baltimore: Johns Hopkins Press, 2015.
Saunders, A. 'Literary Geography: Reforging the Connections'. *Progress in Human Geography*, 34 (4), 2010: 436–452.
Schiller, D. *Digital Capitalism: Networking the Global Market System*. Cambridge, MA: The MIT Press, 2000.
Shelley, M. *Frankenstein; or, The Modern Prometheus*. London: Lackington, Hughes, Harding, Mavor & Jones, 1818.
Smith, H. and Dean, R. T. 'Introduction: Practice-led Research, Practice-led Practice – Towards the Iterative Cyclic Web'. *Practice-led Research, Research-led Practice in the Creative Arts (Research Methods for the Arts and Humanities)*. Eds, Smith, H. and Dean, R. T. Edinburgh: Edinburgh University Press, 2009, 1–38.
Spiegelman, A. *Maus*. New York: Pantheon Books, 1991.
Srnicek, N. *Platform Capitalism*. Cambridge: Polity Press, 2017.

Tabbi, J. 'Introduction'. *The Bloomsbury Handbook of Electronic Literature*. Ed, Tabbi, J. London: Bloomsbury, 2018, 1–9.

Trend, D. *Worlding: Identity, Media and Imagination in a Digital Age*. London: Routledge, 2016.

Tringham, R. 'Creating Narratives of the Past as Recombinant Histories'. *Subjects and Narratives in Archaeology*. Eds, van Dyke, R. M. and Bernbeck, R. Boulder, Colorado: University Press of Colorado, 2015, 27–54.

Walsh, J. *Seed*. London: Editions At Play, 2017.

Wark, M. *A Hacker Manifesto*. Cambridge, Mass.: Harvard University Press, 2004.

Wark. M. *Molecular Red: Theory for the Anthropocene*. London: Verso, 2015.

Wark, M. 'The Vectoralist Class'. *Supercommunity*, 84, August 29 (2015). Web. 10 February 2018. http://supercommunity.e-flux.com/texts/the-vectoralist-class/.

Welfens, P. J. J. *An Accidental Brexit: New EU and Transatlantic Economic Perspectives*. London: Palgrave, 2017.

Part 1

Pasts and presents

Sheds, labyrinths and string figures

2 Creativity today

The case for storytelling

Introduction

In China Miéville's novel *The City & the City* (2009), the twin cities of Besźel and Ul Qoma overlap and intertwine, occupying much of the same space. Yet these cities are also built on the fabled existence of a third city, the mystical Orciny, located in the lost spaces between Besźel and Ul Qoma. This sense of hidden spaces, of a geography culturally and politically obliterated, eviscerated from the conscious mind, echoes across Miéville's work, including his neo-Ballardian short story, 'Reports of Certain Events in London' (2005). For Miéville, space is anything but a simplistic backdrop as his overtly psychogeographic work, *London's Overthrow* (2012), demonstrates. Rather, spatiality is a social-cultural entity, an embodied performance, with omissions, silences and memory through which narratives are either legitimated or closed down.

These theoretical concerns also resonate across the digital narrative, *These Pages Fall Like Ash* (2013) by Tom Abba and Duncan Speakman. The work consists of two elements, a hardcopy book and a digital text, accessed through a smartphone. Both elements can be seen in Figure 2.1.

The physical book consists of two notebooks, each with their own map of a city. One of the cities is Bristol, UK; the other, a parallel city, existing in the same space and time. Locations exist in both worlds. The reader reads the notebooks but also physically explores the shared locations across Bristol, recording their own experiences as they go. At these locations further digital texts are delivered through a smartphone app. As the blurb for *These Pages Fall Like Ash* outlines:

> The story is about a moment when two cities overlap. They exist in the same space and time, but they aren't aware of each other. It's a tale about two people who have become separated, one in each world, about their fading memory of each other and their struggle to reconnect. One of the cities is your own; you become part of the narrative as you travel, moving from place to place. Your version of the story becomes about you and your place in your own city: what would you hold on to? What would you fight to remember?
>
> (2013)

Figure 2.1 Screenshot showing *These Pages Fall Like Ash* by Tom Abba and Duncan Speakman (2013).

Source: © Circumstance (http://wearecircumstance.com/these-pages-fall-like-ash).

Digital stories such as *These Pages Fall Like Ash* are clearly inhabiting a very different creative space to first-generation hypertextual narratives such as Shelley Jackson's *Patchwork Girl* (1995). While the latter is static and hermeneutically enclosed, *These Pages Fall Like Ash* is a performative, affective experience, embracing both a situated and embodied understanding of storifying. It is also transmedial in its use of digital and non-digital elements; indeed, one could say that the digitality of the text arises from the interplay between *both* elements. While *Patchwork Girl* is a meditation on the primacy of the digital (hypertextual) over the physical, *These Pages Fall Like Ash* offers a more nuanced, less demarcated, ontology in which the digital/non-digital divide is porous and artificial.

The growing power and capability of technology underpins this change. Yet it would be wrong to see the transformation in digital storytelling represented by *Patchwork Girl* and *These Pages Fall Like Ash* as purely technological. It is argued in both this and the following chapter that these two stories capture a far-more fundamental shift in the nature of creativity itself. If we are to understand works of digital art such as *These Pages Fall Like Ash* then it is necessary we understand this transformative shift in the modality of creativity. As Harriet Hawkins notes, 'creativity is an idea with much promise, but many tensions' (2017, 338). Its role and function both within the arts and humanities, but also, wider, across academia more generally, is not necessarily straightforward or readily understood. This chapter therefore situates its analysis of creativity within an analysis of practice-based research across the arts and humanities, exploring both the poetics and praxis of, what Tabbi calls, 'our present era of post-digital normativity' (2018, 8).

Creativity: mirrors, lamps and labyrinths

Creativity is a beguilingly simple word. And yet the more you look, the more complex it becomes. Writing at the very beginning of the twenty-first century, Rob Pope noted that 'creativity' as a term had dropped out of favour within literary and cultural theory (2005, 8). Marxist ideology stressed 'production' at the expense of 'creativity' while poststructuralists such as Roland Barthes hailed the death of the author (1967). Yet Pope is not alone when he notes a contemporary return to what he calls 'freshly charged notions of creativity' (2005, 10). David Bohm, for example, stresses the role of creativity as a fundamental essence of human activity. Indeed, Bohm states that 'to awaken the creative state of mind' (1996, 29) is without doubt 'the most important thing to be done in the circumstances in which humanity now finds itself' (1996, 30). George Steiner, in *Grammars of Creation* (2001), argues in the same vein, declaring that ongoing technological innovations are radically changing the concept of individual creation and poetic and philosophical invention.

In trying to outline, at the start of the new millennium, where all this left creativity, Pope came up with the following definition: 'the capacity to make, do or become something fresh and valuable with respect to others as well as ourselves' (2005, xvi). Pope states that he used the term 'make, do or become' because it embraces the different outcomes of creativity: an object, an action or an ongoing process. In other words, the outcome of a creative act may not necessarily be a completed object or performance but something in the process 'of becoming'. Pope goes on to explain that 'fresh' incorporates the ideas of appropriation and intertextuality, recognising that creativity is not simply reserved for the creation of what is considered 'new' but should also reflect radical forms of *re*-creation. Finally, 'with respect to others' captures the central idea that creativity always occurs in relation to 'other people and other things' (2005, xvi–xvii).

It's a useful starting point. However, there's a real danger that such descriptions, in attempting to be universal, are so bland and generalist as to be of no practical value. It's important to remember that this chapter is interested not so much in a discussion of creativity *per se*, but rather in understanding the role and function of creativity within academia, and in particular, as it pertains to the arts and humanities disciplines. Indeed, Bohm's own warning that 'creativity is, in my view, something that it is impossible to define in words' (1996, 1), should be enough to alert us to the absolute necessity of a tight critical focus, whatever that might mean. After all, at its broadest definition, creativity is not just the production of physical artefacts or performances; as I've argued already, it is also an inherent part of cognitive psychology, the ongoing sense-making of our brains that is called living.

Once we start to narrow our investigation in this way, a number of things begin to happen. The first is that we start to get a sense of what creativity is for, and then, with this clarity, comes a firmer sense of what is actually being produced (the creative artefact) and the process that led to its making. In academia, it could be argued that creativity runs across everything that counts as research – from the writing of academic (factual) books and articles, through to physical (quantitative) experiments. Each clearly involves a creative engagement with a discipline of some kind. Yet the fact that terms such as 'creative turn' and 'creative writing' are not regarded as tautological are reminders that the adjective *creative* and noun *creativity* have very particular meanings and associations within academia, especially when it comes to research. Put simply, creativity is associated with the production of a 'creative artefact'. This artefact could be a physical thing such as a novel, poem or painting; or it could be a performance. But what it won't be is a factual book or article, despite the fact that, in their own way, in their written form and construction of argument for example, they too are a form of creativity. In other words, creativity is seen as a distinct methodology within academia from which arises a distinct output, distinct in the sense that, whatever it is, it will not be a factual (thesis-driven) book or article. Inscribed here then is the classic division between traditional, non-creative research outputs (books, articles, papers) and the non-traditional, creative output (poem, novel, performance). It is this understanding of creativity that operates predominantly within academia, and it is this sense of creativity which will be used as a starting point for this book (see Smith and Dean 2009; Barrett and Bolt 2010). Note, however, that this definition of creativity is not necessarily domain or discipline specific. Although creativity is very much associated with the arts and humanities, it could equally be used across the sciences, as long as the output conforms to the sort of creative artefact described here (for example, see Ulrike Felt *et al.*, *The Handbook of Science and Technology Studies* 2017). I will say more about this when I discuss practice-based research in the next chapter, particularly the intimate relationship between 'creativity' and 'criticality' (Barrett and Bolt 2010). For now though it's enough to recognise the focus that this book has in terms of its analysis of creativity: the meanings and associations that creativity has within academia.

There's something else that needs consideration too. Both Pope and Kearney recognise that creativity has not been an historically inert concept. In his book, *The Wake of Imagination* (1988), Richard Kearney outlines what he understands as the paradigm shifts in Western European artistic representation. He isolates three separate shifts or paradigms which each privileged a separate 'metaphor characterizing the dominant function of imagination' at that time (1988, 17): he labels these mirror, lamp and labyrinth. In the first period, associated with classical thought, creativity is presented as a (mimetic) *mirror* that reflects reality; in the second, associated with the Romantic period, creativity is understood as an (expressivist) *lamp* that generates its own heat and light; and in the third, the postmodern imagination is understood as a (virtual) *labyrinth* of looking glasses that reflect infinite variations on an ultimately illusory object (1988, 15–16). Although only a crude model, it reminds us of at least two important points. First, that creativity is not an objective, free-floating concept but is tethered to something far-more fundamental: ontology. In other words, creativity is but one element of far deeper issues concerning how the world is seen and understood. And second, that this ontology, this way of seeing and understanding reality, has changed, and will continue to change over time.

In postmodern art, the labyrinth becomes a metaphor of the self from which the outside world is unknowable (Harvey 1990). The labyrinth is also one of language. The postmodern artist cannot overcome the inherent arbitrariness that lies behind representation and its meaning.

Yet, as we'll see in this chapter, it is now readily accepted that this era of postmodernity is over (for example, Moraru 2011). But if we have indeed left the labyrinth of mirrors then where exactly are we? What is the 'new cultural imaginary', to use Moraru's phrase? These are questions whose answers are only now emerging, yet they are questions this and the following chapter address head on. Whatever direction we're headed in, Pope writes that the theory and practice of creativity must try to reach beyond the limitations of academia, with its rigid divisions between arts, sciences and technology (2005, 22). He concludes, quoting the Nigerian writer Chinua Achebe, 'people create stories create people create stories … we live by "narratives" ("grand" or "small") and they, in turn, live and evolve through us' (2005, 191).

Hawkins pushes this further. In her study, she lists three 'critical geographies' of a new kind of creativity, a creativity for a post-postmodern age. The first is creativity as an embodied, material and social practice; the second is the politics of creativity; and the third is what Hawkins terms 'creativity as a force in the world' (2017, 12). Each of these categories offers ways in which Pope's original definition might be reconceptualised for a new age of creative modality, particularly in regard to our understanding of creativity as a research methodology. Hawkins' first 'critical geography' is perhaps the most interesting, foregrounding the central importance of creativity as a situated and embodied act. In other words, creativity takes place in a particular place and time by a particular human being. As Hawkins goes on to say,

> [Creativity] is a practice found in specialist spaces of the studio and the museum, but also in the home and on the streets, it is a practice wrought from decades of skilled learning and bodily habituation, as well as an unconscious in the moment act of improvisation or just part of what it means to be in the world.
>
> (2017, 338)

This, 'what it means to be in the world', will become increasingly important as I turn to look at Heidegger's notion of 'being-in-the-world'. For now though it's enough to recognise the emphasis that Hawkins is placing on creativity as a *social practice*. In other words, it emerges through networks and physical connections; it is through such networks that what is understood by creativity is itself constantly negotiated.

Hawkins' second critical geography is that of the political landscape through which any discourse of creativity arises. This landscape permeates and affects all levels of the creative process. Creativity relies heavily on government funding, either directly in the commissioning of work, or indirectly in the running of museums, independent galleries and art studios. As such, it can be particularly vulnerable to the vagaries of government policy and political ideology. Politics also infuses our understanding of identity, community and what might be understood as subcultural practices. The social networks mentioned above are also inherently politicised, of course, operating as a politically-charged discourse through which ideas and concepts are legitimated. For some, creativity has become a handmaiden of neoliberalism, associated with governmental efforts to try and regenerate urban economies through the promotion of creative industries (Peck 2009). Yet, as we'll see, creativity can also offer the means of articulating alternative modes of thought, both transgressive and resistant to social or economic norms. As Hawkins says, 'creative practices not only offer modes of intervention, artistic or otherwise, but also establish ways to live differently' (2017, 20). This resistance does not come from outside of the neoliberal/capitalism machine but rather operates from within it.

Hawkins' third category seeks to understand creativity as an affective force in the world. In other words, Hawkins represents creativity as something that is inherently transformational and dynamic. As she states, 'creative practices … are viewed as making place, as forming and transforming the subjects who practice and engage with these practices, and shaping forms of knowledge production' (2017, 22). In a world of increasing social, economic and political complexity, creativity is therefore seen as offering the means to engage, challenge and transform.

For Kearney, it is this *ethical* underpinning of the new creative paradigm that remains paramount, clearly separating it from the postmodern irony and depthlessness of what went before. In his analysis of historical narratives, Kearney lists three ways in which creative storytelling can assist in the process of what he calls 'empathic identification' with the past (1998, 255). The first is what he calls the *testimonial capacity* of creative storytelling to bear witness to a forgotten past; the second, the *empathic capacity* of storytelling which allows us to identify with those who are different to us; and lastly, the *critical-utopian capacity* of storytelling by which we can challenge official stories 'which open up alternative ways of being' (1998, 255). As Kearney goes on to say:

> While narrative imagination is … not always on the side of the angels, it does possess the power to disclose dimensions of otherness. And it is ultimately this power of disclosure which marks the basic ethical ability to imagine oneself as another.
>
> (1998, 255)

Although Kearney is discussing historical fiction, I would argue that his concept of ethical responsibility runs across all creative practice within academia. If creativity is to be an affective force in the world, Hawkins' third category, then it needs to engage with the most pressing global issues facing us, including (but certainly not limited to) ecological, urban and social disruption.

Let's look at an example of this. Anne Brewster has devised, what she terms, 'personally-situated writing' to engage with issues of whiteness and Indigenous sovereignty in Australia. For Brewster, personally-situated writing foregrounds the embodied nature of cultural norms, especially in regards to race and power dynamics. She draws on the work of Weiss and her term 'embodied ethics', highlighting the importance of bodily (situated) imperatives within the ethical agency of personally-situated writing (2009, 129).

> The beach is a meeting place. The waves tumble. They are saying something but it is in a language we never knew. Behind, our footsteps fill with water. Around us the waves remake the beach. At the mouth of the river stand five black men, watching.
>
> (2009, 142)

The writing merges form and style, recombining fiction with non-fiction, a bricolage of 'journal, anecdote, poetry and the novel' (2009, 142) reminiscent of Ruth Tringham's 'recombinant histories' discussed in Chapter 1. At all times the experience is personal and embodied. As Brewster explains, the author of personally-situated writing 'is not an inert, passive observer but one who is embedded in the fraught standoff between white and indigenous sovereignties that subtends the white nation' (2009, 135). In other words, through such writing, the cultural hegemony of whiteness is itself defamiliarised from within.

At least two questions follow on from this: first, where does that leave us with Pope's original description of creativity; and, second, if there has been any kind of change or modification in how creativity is understood, then what does that start to say about the, as yet, unidentified fourth stage in the 'mirror, lamps and labyrinths' schematic described earlier (Kearney 1988).

When considering the first question, it's important to remember that, though this book's focus is storytelling, one of its overarching themes remains the uses of creativity within the arts and humanities. And it is from this perspective that I'd like to engage with Pope's definition. Here it is again: 'the capacity to make, do or become something fresh and valuable with respect to others as well as ourselves' (2005, xvi).

In my discussion of the work of Hawkins, Kearney and Brewster, I've indicated some key areas where this definition might be considered lacking, particularly in regards to the arts and humanities. These can be broken down into three distinct positions that are aligned with Hawkins' 'critical geographies'. The first is the centrality of creativity to how individuals, communities and nations understand, engage and co-create the world. In other words, creativity is not just the pastime of those with the time and money to consider themselves artistically inclined. At its very heart, creativity is about something far-more fundamental: knowledge creation. Creativity is therefore not simply a mirror held up to the world; it also offers a means of affecting transformative change and adaptation, or, to use Hawkins' words, 'to research and live differently' (2017, 22). Creativity as an activity therefore comes with ethical responsibility. Second, creativity is both a situated and an embodied activity. Creativity does not occur in a vacuum but has both spatial and temporal dimensions. It is also undertaken by a creator or creators, human beings caught in the specifics of their own, and others, lives. As Brewster says of personally-situated writing, it allows her to 'not merely "think" my way through a problem or issue, but immerse myself in its bodily and affective dimensions' (2009, 142). Yet creativity is not merely embodied and situational, it is also a social practice: it both creates and emerges from social and cultural networks. Third, creativity is an inherently political activity. An example of this would be the way that creativity has become part of the neoliberal discourse around economic regeneration (Peck 2009). UNESCO's Creative Cities Network was created in 2004 with the express goal of placing creative and cultural industries at the heart of urban growth and development. Government policy and ideology directly influence academic funding streams which in turn have a direct effect on creative projects and initiatives. It is also political in the sense that those creative activities outside of this neoliberal/capitalist consensus – activities that may be called radical, avant-garde or subversive – are by their very difference also inherently politicised. Donna J. Haraway uses the word *trouble* to describe this sort of thing, a word which, in its original thirteenth-century French, meant 'to stir up', 'to make cloudy' and 'to disturb' (2016, 1). In fact, in the face of the Anthropocene, Haraway argues it is only by creating trouble that we will continue to exist as a species: 'our task is to make trouble, to stir up potent response to devastating events' (2016, 1). Stories and storytelling are central to Haraway's notion of trouble, a position underlined by Bruno Latour in his essay, *Down to Earth: Politics in the New Climatic Regime* (2018).

As I've said, these three issues – creativity as knowledge creation, as situated/embodied and as politicised – take on even more importance when creativity is understood as a research paradigm from the perspective of the arts and humanities disciplines. Keri Facer and Kate Pahl, in their analysis of the Connected Communities Programme, provide a useful insight into this. Funded by Research Councils UK and the Arts and Humanities Research Council (AHRC), the Connected Communities Programme was established with the aim of funding projects, partnerships and networks that would 'create a deeper and richer understanding of "communities" in all their historic and contemporary forms' (Facer and Pahl 2017, 9). The programme has at its heart both a *collaborative approach* to arts and humanities research – in other words, it self-consciously seeks to break down the boundaries between academia and the wider world, including citizens, public bodies and community organisations – and an *interdisciplinary approach* in which projects are encouraged to bring arts and humanities methods and theories into dialogue with other disciplines and scholarship. In their review of over 300 projects, involving over 500 collaborating organisations and 700 UK academics, Facer and Pahl list what they see as the key elements of these projects (2017, 17). This includes: (1) *Materiality and place* – the fundamental iterative relationship between knowledge production and objects, landscapes and cultures; (2) *Praxis* – knowledge through doing, what Facer and Pahl call a 'performative ontology that shapes the world as it studies it'; (3) *Stories* – the role narrative plays in the creation of knowledge; and (4) *Embodied learning* – that knowledge is fundamentally embodied and situated, whether individually or/and through embodied networks and communities (2017, 17).

Preserving Place: A Cultural Mapping Exercise is a good example of this, a project funded by the AHRC under the Connected Communities Programme (Smyth *et al.* 2017). The aim of the project was to explore how cultural mapping can be used to record and understand the legacy of collaborative heritage research (2017, 191). Karen Smyth *et al.* quote from the Cultural Mapping Toolkit to help define what they mean:

> Cultural mapping allows us to see where we've been and where we are in order to find our way forward, just as any mapping process might. The difference is the objects of cultural mapping are not topographical features, but tangibles like assets and resources and intangibles like identity, relationships and possibilities.
>
> (2017, 70)

The project used quantitative questionnaires and qualitative interviews to capture the histories, processes and everyday experiences of various heritage groups (2017, 200). An online mapping tool was then developed to visually represent the project timeline, providing both a dynamic interface for project planning as well as an 'asset map', allowing access to quotes and commentaries, as well as full-version stories. The four categories listed above (materiality and place, praxis, stories and embedded learning) remain underlying features of the methodology: the project used a cultural mapping methodology in which place and practice were intimately linked; it foregrounded both praxis, in terms of how narratives were collected and brought together, and storytelling as a key aspect of the research; and it prioritised embedded learning, explicitly seeking to map the situated nature of knowledge construction across the heritage community. As the Cultural Mapping Toolkit extols, 'mapping our cultures in all their 360 degrees and in their depth, keeping our diverse stories and multiple histories interrogating and renewing each other, is a key cultural function of our time' (2017, 25).

Maps, string and rhizomes

Both Brewster's personally-situated writing project and Smyth's cultural mapping project are just two examples of how creativity is now being used as a fundamental part of research within the arts and humanities. The first is primarily an individual (personalised) activity, where the creativity is undertaken through the composition of creative writing. However, in the second, the range of creativity is far broader and less straightforward. Stories and narratives are told by participants in what is a creative (narrative-led) process; but then these stories are put together by the project team as part of the mapping process; in other words they are abstracted, edited and arranged into a kind of meta narrative (or map) through which any user then has to navigate and explore. This is clearly a creative (narrative-led or storifying) process in itself. The writer and academic Peter Turchi is not alone in recognising that maps are indeed a kind of story in and of themselves: 'To ask for a map is to say, "Tell me a story"' (2004, 11). If a map is also a story then it too needs an author or authors. Conversely, he recognises that 'a story or novel is a kind of map because, like a map, it is not a world, but it evokes one' (2004, 166).

A map seems a particularly powerful metaphor by which to think about the way creativity works. If creative artefacts are, at least in some way, offering a kind of story or narrative, even if they are not actually textual in nature, then a map seems one way in which they might be theorised. In this sense, a map is understood as a symbolic representation (signifier) of something else, a creative abstraction shaped by form, context and convention (Turchi 2004, 25; see also Ryan, Foote and Azaryahu 2016). A good example of this is given in Figure 2.2 which shows the oldest original cartographic artefact in the Library of Congress, a portolan nautical chart of the Mediterranean Sea, made sometime in the early fourteenth century. Portolan charts were navigational maps made by sailors taking their own compass directions and estimated distances from their ship. Figure 2.2 shows the rhumbline networks that characterise this sort of chart. The lines are designated lines of bearing made through observation and compass measurement, allowing those using the chart to set a course from one harbour to another.

Hawkins' critical geographies are equally applicable to a map too. The production of a map involves explicit knowledge creation, the cartographic act itself bringing forth new worlds and horizons; map-making is situated and embodied, its meaning and intent embedded within the context of its creation; and, lastly, map-making is a highly politicised act, the agent both of hegemony and oppression as well as subversion and counter-culterancy, as Jorge Luis Borges' short story, 'On Exactitude In Science' (1946) reminds

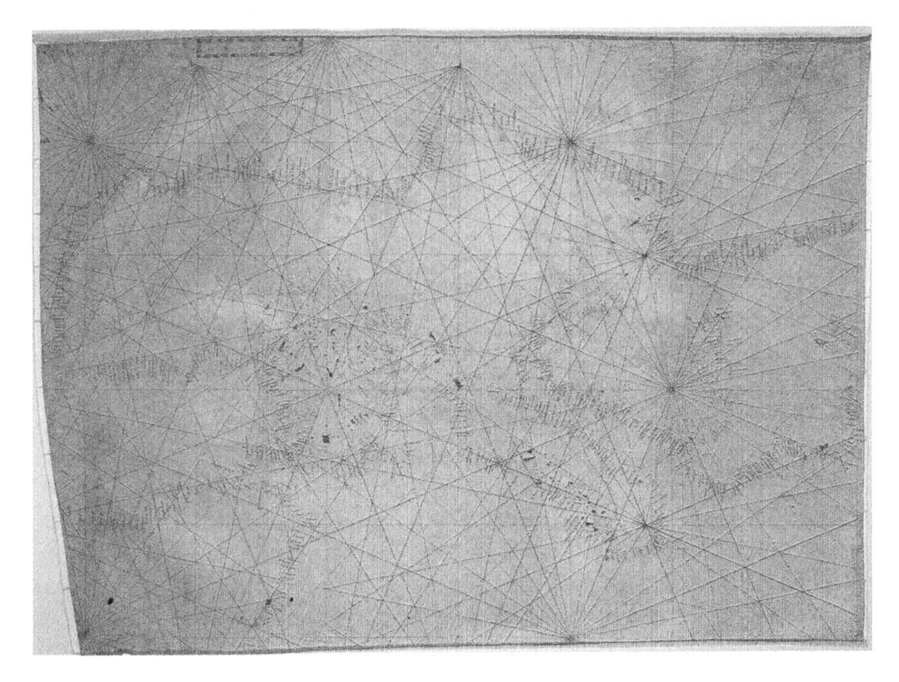

Figure 2.2 Portolan chart of the Mediterranean Sea ca. 1320–1350: manuscript chart of the Mediterranean and Black Seas on vellum, 43 × 59 cm.

Source: © Library of Congress, Geography and Map Division.

us. And if creative output is equivalent to a map, then creativity itself is a kind of map-making – a way-finding, if you will, through an existence with no inherent paths or sense. I will say more about this later on; for now it's enough to recognise that the metaphor of map-making offers at least something in regard to my ongoing consideration of our new creativity modality.

But map-making is not the only metaphor in town. Perhaps a less familiar one that I feel has particular traction in terms of this discussion is Haraway's use of string figures (2016, 9–16). String figures are the shapes created by the manipulation of string around the fingers. They can be created individually or can involve multiple people. Perhaps the most famous use of string figures is in the game known as cat's cradle. Here, two or more players take it in turns to manipulate the string pattern, each altering the figure made by the previous player. According to Haraway, string figures are among humanity's oldest games (2016, 13). They appear all over the world and are used to tell stories, as divination and as play. Haraway explains how, for the native American Navajo people, string games are just one form of what she calls 'continuous weaving', in other words, 'practices for telling the stories of the constellations, of the emergence of the People, of the Diné' (2016, 14). Yet they also tell some of the newest stories too: in string theory, a theoretical framework of particle physics, strings propagate, intertwine and vibrate at the subatomic level of all physical matter (Green 2000). As Haraway goes on to say:

> Playing games of string figures is about giving and receiving patterns, dropping threads and failing but sometimes finding something that works, something consequential and maybe even beautiful, that wasn't there before, of relaying connections that matter, of telling stories in hand upon hand, digit upon digit ...
>
> (2016, 10)

Crucially, then, string figures are thinking as much as making practices (Haraway 2016, 14). In Vera Frankel's innovative artwork from 1974, *String Games: Improvisations for Inter-city Video*, two groups of people played a game of cat's cradle between Montreal and Toronto, the groups connected through an early form of teleconference transmission (Tuer 2006, 38). According to Tuer, the poignancy of the participant's gestures, linked and intertwined, physically through the string and digitally through interconnecting lines of telecommunication, was a key feature of the performance, 'their desire to reach across time and space to one another central to the exploration of new technologies in Frankel's work' (2006, 38). Figure 2.3 shows a selection of street rehearsal shots from the artwork.

One could consider Frankel's work a creative extension of the tin-can telephone, the sort of thing children experiment with in the garden or on the street, shouting into two cans connected by a taut length of string. From here it's just a short technological jump to the cat's cradle of the switchboard, with their complex interlacing of hands, fingers and cords, a technology we'll be meeting again in Chapter 4 (Figure 2.4).

Unlike the metaphor of map-making, which emphasises the *situatedness* of creativity, its constructiveness and abstraction, string figures highlight the *interconnectedness* of creativity, its contingency and tactility. This is creativity in real-time, the world transmuted into endless sequences of symbolic patterns and voices, a code devised and transmitted from finger to finger, hand to hand 'in twists and skeins ... holding still and moving, anchoring and launching' (Haraway 2016, 10).

As I've already said, Haraway's project is to stir up 'trouble', to affect a transformative change in the way we think about the world in the face of mounting crises. For Haraway, governmental responses to impending ecological disaster are inadequate and overly bureaucratic. Instead 'revolt needs other forms of action and other stories for solace, inspiration, and effectiveness' (2016, 49). It is in this context that Haraway sees the metaphor of the string figure (individual and collective, shared and co-created) as being so powerful and enabling. For her, string games are metaphors for a new kind of storytelling; a new way of engaging and ultimately changing the world. Yet I would go further than this. I would say that string figures are a powerful metaphor for creativity *per se*. In short, string figures and games capture the

Figure 2.3 *String Games: Improvisations for Inter-City Video* (Montreal-Toronto, 1974) by Vera Frenkel.

Source: © Courtesy of Vera Frenkel.

Figure 2.4 Manual assistance operators at work in the Nambour Telephone Exchange, 1966.

Source: © Picture Sunshine Coast, Sunshine Coast Council.

contemporary imperatives that underpin our creative and ethical entanglement with the world.

Before I go on, let's just remind ourselves what I've been arguing here. So far this chapter has been focussed on defining what creativity actually is today. I've taken two starting points in this debate. The first is Pope's definition outlined at the start of his book, *Creativity: Theory, History, Practice* (2005): 'the capacity to make, do or become something fresh and valuable with respect to others as well as ourselves' (xvi). The second is Kearney's historical schematic of the paradigm shifts in Western European artistic representation, summarised as mirror, lamp and labyrinth (1988). Pope's definition offers a rhetorical straw man, a device through which we might begin to piece together a revised, even radical, reassessment of creativity. Kearney's schematic helps here, in the sense that it points to the footprints that have already been made in the sand while at the same time inviting discussion of what kind of paradigm, beyond the postmodern labyrinth, we are now inhabiting.

I've also done something else in this chapter. I've made it clear that the book is focussed on creativity as it pertains to academia, and most especially the arts and humanities. In particular I'm interested in how creativity can be understood as a form of research of and in itself. By clarifying this delineation I am making clear that my understanding of creativity embraces the production and research of (non-traditional) creative/performative output, that is defined, at least in part, by what it is not: the (traditional) factual academic thesis, essay or book. This is not to ignore the complex and disputed nature of this division: the relationship between 'the creative' and 'the critical' is certainly 'intimate and shifting' (Pope 2005, xvii). I'm going to be saying a lot more about this further on in the chapter when I turn to practice-based research. But for now though I'd simply like to recognise this focus.

With these parameters in place, I then put forward two different ways of thinking about how we might conceptualise creativity in this new modality: one is as a form of map-making; the other is as a string game, through the creation of string figures. Pope has a term for these kind of metaphors, descriptive models that help us envisage and theorise complex phenomenon: he calls them 'images for imagining' (2005, 17). Yet, however they might be described, the metaphors outlined here provide a powerful way of rethinking what creativity is today, offering a means of instantiating a transformative 'way of becoming' (Pope 2005, 17). Both emphasise the situatedness and embodied nature of creativity. Both foreground its social and communal aspects. And finally both exhibit the sort of ethical and political imperatives that Hawkins, Brewster and Haraway, amongst many others, argue must now be a fundamental aspect of creativity. Clearly, this is not the anodyne creativity of neoliberal governmental policy, the dogma of industrial productivity and market growth. Instead, it is the kind of creativity that comes from a sense of ethical, political and social responsibility. It is the sort of creativity that springs from the desire to do the sort of 'troublemaking' that Haraway prescribes; that reaches out to the empathic and critical-utopian approaches outlined by Kearney (1998) and that engenders the sort of 'othering' that Pope calls becoming 'other-wise' (2005, 29).

It would be tempting to limit this analysis to these two models. Yet I can't conclude this section without serious consideration of perhaps one of the most powerful metaphors in poststructuralist thought, that of the rhizome. In their book, *What is Philosophy?* (1991), Gilles Deleuze and Félix Guattari outline, what they see as, the fundamental role of creativity within philosophy. They go on to describe three intersecting 'domains' in which creativity occurs: the first is *philosophy*, which is involved in the creation of concepts; the second is *art*, including literature, which is involved in the creation of (sensory) affects; and the third is science, which involves the creation of *precepts* (1991, Chapter 7). Crucially, Deleuze and Guattari emphasise the interconnectedness of each domain, in that each is a creative mode of thought, seeking to bring order and structure to 'chaos': 'the three modes of thought intersect and intertwine' such that 'a rich tissue of correspondences can be established between the planes' (1991, 198–199). One purpose of their book was to rescue creativity as a concept from the claws of capitalism by stressing creativity's radical and transformative potential, as opposed to its subservience to any post-industrial economy. For Deleuze and Guattari, then, it was essential that creativity operated free of imposed power structures. In *A Thousand Plateaus* (1988) Deleuze and Guattari use the metaphor of the (arboreal) tree to signify this traditional hegemony, a logical hierarchy of root, trunk, branch, twig and bud, all ordered and firmly rooted in its place. As an alternative 'image of thought' (Deleuze 1968), free of such formal rigidity, Deleuze and Guattari offer the example of the rhizome (such as ginger, asparagus and irises). While roots extend downwards, pulling up moisture and nutrients, rhizomes are underground stems. As such, they tend to grow horizontally, just under the soil, sprouting roots and shooting up new vertical stems as they go. They can also consist of interdependent biological systems such as viruses, orchids and wasps.

Figure 2.5 shows Richard Giblett's artistic representation of an interconnected, cross-species rhizomatic structure with the mycelium of fungi. It is an artistic abstraction only, of course, yet at its heart is the nodal, subterranean, non-hierarchical structure that Deleuze and Guattari prescribe. As they state, a 'rhizome has no beginning or end; it is always in the middle, between things, interbeing, intermezzo' (1988, 25).

The rhizome then is a non-hierarchical network by which creative thinking can be liberated from the arboreal condition of monolithic authoritarianism. The distributed structure of the rhizome that we see in Figure 2.5 makes it resilient, adaptable and anarchic. It is democratic, communal, bottom-up and transgressive. It is a trouble-making string figure hidden below ground. It is cross-species. Indeed, for Haraway, the only way humankind can survive is by embracing a multispecies 'sympoiesis' or 'making-with' (2016, 58): 'We are humus, not Homo,

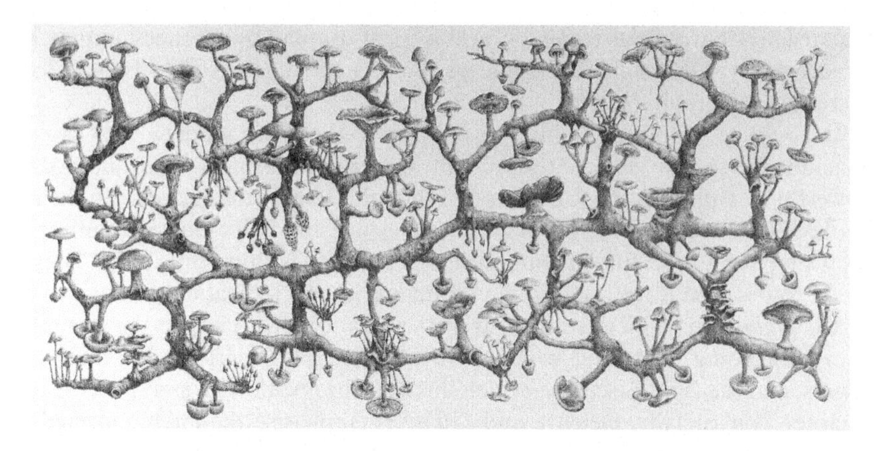

Figure 2.5 Richard Giblett 1966–2018, *Mycelium Rhizome* (2009), pencil on paper 120 × 240 cm, private collection.

Source: Courtesy of the Richard Giblett Estate and Murray White Room, Melbourne.

not anthropos; we are compost, not posthuman' (2016, 55). Yet if the rhizome support's Haraway's notion of multispecies sympoiesis, then it can also be understood as a form of map-making too. As Deleuze and Guattari state, the rhizome 'pertains to a map that must be produced, constructed, a map that is always detachable, connectable, reversible, modifiable, and has multiple entranceways and exits and its own lines of flight' (1988, 23). Like the rhizome itself, hidden in its earthly compost, these three 'images of imagining' of the map, string figure and rhizome, intertwine and co-adapt.

So where does the map, the string figure and the rhizome leave us in terms of our understanding of creativity today? I would argue that each supports a way of thinking about creativity that allows us to enhance Pope's original definition. That definition, of course, was a very general one, and purposely so. Yet it serves a useful function here in thinking about the focus and limits of creativity as they apply to this book before we drill down into a detailed analysis of storytelling. Here is Pope's definition one more time: 'the capacity to make, do or become something fresh and valuable with respect to others as well as ourselves' (2005, xvi).

As I've already stated, this book is primarily interested in creativity as it applies to the arts and humanities disciplines within academia. As such, I would argue that any notion of creativity must embrace the following:

1. An understanding of creativity as a fundamental process of *knowledge creation*. Creativity is central to our engagement with, and co-creation of, materiality, place, and selfhood;
2. An understanding of creativity as an inherently *situated* and *embodied* activity;

3. An understanding of creativity as a means of '*becoming other-wise*'. By this I mean that creativity as practiced in academia cannot afford to be solipsistic, in and of itself. It must seek to affect change; to cause trouble, as Haraway would have it, to explore Kearney's 'dimensions of otherness' (1998, 255).

Taking these three characteristics, it is possible to refashion a new description of creativity that we can take forward.

> Creativity is a situated and embodied act, in which, through the (co-) creation of new knowledge, perceptions and understanding of the world are changed.

In this description, the social-political context of the creative act is foregrounded, as is its transformational (affective) intent. The emphasis on the creation of new knowledge, and through that, a concomitant change in understanding, is therefore critical, superseding the rather innocuous 'fresh' and 'valuable' of Pope's original definition. The recognition of co-creation is important too. One of the things we've seen in this chapter so far is the possibility that creativity can be a collaborative or/and shared experience. Indeed, one of the key features of both the rhizome and the string figure is that they extol a complex lattice-work of (cross-domain) interdependency.

With this revised understanding of creativity, it is now possible to revisit Kearney's mirror, lamp and labyrinth schematic of artistic representation (1988). As I've already argued, it is now clear that, although postmodern theory has left a lasting legacy on artistic representation, we are indeed entering a new paradigm. As Alison Gibbons notes, 'a new dominant cultural logic is emerging; the world – or in any case, the literary cosmos – is rearranging itself' ('Postmodernism is Dead'). For Robin van den Akker, Alison Gibbons and Timotheus Vermeulen, this new era emerged in the 2000s (2017, 4). The fall of the Berlin Wall in 1989, the 9/11 attacks and the financial crisis have all been seen as key indicators of a paradigm shift in how western capitalist society is understood (2017, 4). To this list one could also add the election of Donald Trump as President of the United States of America; the pro-Brexit vote in the UK referendum and the rise of populist politics around the world; and the growing awareness of the Anthropocene as a period of unprecedented ecological threat. In *Dead Pledges* (2017) Annie McClanahan notes that 'this sense of crisis has become both the ambient context and the manifest content of cultural production' (15). Although McClanahan is referring to the systemic effects of global debt after the financial crash, her comments resonate strongly with broader concerns of crisis as a key twenty-first century condition.

As neoliberalism has faltered then so too has its alter ego, postmodernism. The certainties of Left and Right have disappeared. The self-referential and solipsistic focus of postmodernism seems, at best, quaint and old fashioned and, at worst, ethically and morally irresponsible at a time when the continued existence of our, and many other, species are under threat. It is such moral and ethical connectedness that remains a key feature of this new era. van den Akker, Gibbons and Vermeulen, and Mary K. Holland, amongst others, have called this period 'metamodernism', although it has a host of other names too, including post-postmodernism, cosmodernism (Moraru 2011) and even digimodernism (Kirby 2009). While Christian Moraru's cosmodernism calls for 'a new togetherness, for a solidarity across political, ethnic, racial, religious, and other boundaries' (2011, 5), metamodernism emphasises an in-betweenness of feeling, an oscillation between lingering postmodern anxiety and the need for ethical force and truthfulness. As Gibbons notes, this multivalency is neatly reflected in the anxiety around the term 'post truth' itself, a concept that seems to be rejecting truth as an ontological possibility at the same time that it is reaching out for some new kind of consensus (and hope) ('Postmodernism is Dead').

These ideas resonate across this book. Although metamodernism is still a tentative, even controversial term, I would argue that it provides a useful theoretical position by which to deepen and extend our discussions of contemporary creativity. This is not too deny that alternative approaches such as Moraru's cosmodernism are not without attraction; cosmodernism's relationship to 'worlding of the world' (Moraru 2011, 3), for example, has real relevance to the debates in this, and other, chapters. And digimodernism's focus on, what Kirby calls, 'the effects on cultural forms of digitisation' (2009, 253), is equally pertinent in many ways. Yet I would argue that van den Akker, Gibbons and Vermeulen's discussion of creative process and technique make metamodernism a far-more enabling concept by which to engage with creativity.

In her analysis of Ben Lerner's novel *10:04* (2015), Gibbons notes how the narrator of the story (a first person voice who could or could not be the author) repeatedly intrudes in its telling. Superficially it appears to offer the same kind of self-referentiality that characterises John Fowles' *The French Lieutenant's Woman* (1969), a novel firmly within the postmodern canon. As Mahmoud Salami states of Fowles' novel, 'the self-referentiality that is emphasised throughout the novel suggests that history is a text and can never refer unproblematically to an empirical reality' (1992, 110). Yet, according to Gibbons, the metatextuality in Lerner's *10:04* has a very different effect. Here, the authorial intrusions are less about signalling the inherent fictionalising process of knowledge creation than about recognising an ontological position that is inherently situated and embodied: 'in *10.04*, this rearrangement of the world turns on an axis of human subjectivity, conceived as intimately and ethically relational' (Gibbons 'Postmodernism is Dead').[1] Gibbons states that Lerner's novel can be categorised as 'autofiction', a genre that blends autobiography with fiction. For Gibbons, the growth of autofiction as a literary genre is a key feature of the metamodern condition ('Contemporary Autofiction').

> Contemporary autofictions do not only narrativise the self, but they also thematise the sociological and phenomenological dimensions of personal life such as how identities relate to social roles, how time and space are lived and how experience is often mediated by textual and/or digital communication. It is in this sense that metamodern affect is situational; it is ironic yet sincere, sceptical yet heartfelt, solipsistic yet desiring of connection. Most of all, it is experiential.
>
> ('Contemporary Autofiction' 2017, 130)

The sort of autofictions that Gibbons talks about seem to be very closely allied to Brewster's personally-situated writing. Both provide compelling evidence that postmodernity no longer offers the kind of critical baseline it did in the second quarter of the twentieth century. Kearney's metaphor of the labyrinth (1988) does indeed appear to be have been left behind in the scramble for a new kind of creative engagement. van den Akker, Gibbons and Vermeulen's concept of metamodernism offers a theoretical way forward here, certainly one we can interrogate in this book. Their notion of a creativity that embraces its inherent situatedness, that seeks a new kind of ethical and moral responsibility, that is both sceptical of power and authority, but nevertheless is not afraid to respond to the imperative for a new kind of realism, is one that chimes strongly with what I've been arguing so far. Here's my revised definition of creativity again.

> Creativity is a situated and embodied act, in which, through the (co-) creation of new knowledge, perceptions and understanding of the world are changed.

As a baseline definition of metamodernist creativity, I would suggest it has value, stressing as it does both the situated and embodied nature of the creative act. The emphasis on creativity's ability to change our 'understanding of the world' speaks to the sort of metamodernist embracing of realist and ethical imperatives that van den Akker, Gibbons and Vermeulen describe.

So, if we have indeed entered a new period of artistic representation, how is that best encapsulated? This chapter has presented three possible metaphors that could provide a fourth category in Kearney's schematic: maps, string figures and rhizomes.

As we'll see later on, maps and rhizomes have a complicated history in the sense that both have also been used as metaphors of the postmodern condition. Writing in the early years of the web, Stuart Moulthrop drew on Deleuze and Guattari's concept of the rhizome as a means of theorising the poststructuralist potential of hypertext. The map too has been used as a classic postmodern device. As Jean Baudrillard famously stated, '[t]he territory no longer precedes the map, nor does it survive it. It is nevertheless the map that precedes the territory—precession of simulacra—that engenders the territory' (1994, 1). In postmodernity, we have lost all contact with the real world: all that remains, all that we can call 'real', is the artificiality of the map.

Under postmodernism, then, the map and rhizome were highly effective at representing an ontological position which fundamentally denied the ability to access any notion of an unconstructed, 'truthful', reality. And yet today, while the influence of postmodernism wanes, I would argue that the map and the rhizome remain as powerful metaphors, particularly when it comes to understanding creativity. Rather like metamodernism itself, they have transformed themselves, offering powerful models for the sort of situated and embodied creativity espoused by van den Akker, Gibbons and Vermeulen. In this new era of metamodernity, the rhizome and the map are reconfigured as symbols of our subjectively intimate and ethically-relational connectedness. As Christiane Paul and Malcolm Levy note in their discussion of the New Aesthetic movement, the rhizome can be seen as capturing, what they identify as, the progressive collectivism at the heart of the postdigital hacker ethic. Rather than highlighting the constructiveness of knowledge, the arbitrary relationship between signifier and signified, the map and rhizome embody a radical interconnectedness by which we (personally and collectively) might become other-wise. It is here, in this becoming other-wise, that the map and rhizome capture the ethical and moral interdependencies of our existence.

Haraway's string figures, our third metaphor, is especially useful here. As we've seen, for Haraway, the string figure is a way of representing the essential connectedness of the human species, a connectedness with each other, but also a connectedness that reaches out to other species with whom we share our planet. This is no time for postmodern tricksiness or self-indulgence: '[t]he edge of extinction is not just a metaphor; system collapse is not a thriller. Ask any refugee of any species' (2016, 102). In the deepening social and ecological crises of our new millennium, a belief in human exceptionalism and utilitarian individualism will no longer suffice (2016, 57). Nothing less than a paradigm shift in how we understand the world and address its problems is necessary; and for Haraway this reconfiguration of knowledge begins with storytelling: 'it matters what thoughts think thoughts; it matters what stories tell stories' (2016, 39). The string figure represents this 'ecology of practices' (2016, 34), this interconnectedness, this 'making with' or sympoiesis, this troublemaking. It is the metamodernist trope of map and rhizome taken to its logical extreme – an extreme of oneness and multispecies dependency.

Neither metaphor is wrong or right, of course. Instead each offers a different way of conceptualising creativity today, particularly within the arts and humanities. At the core of this conceptualisation, however, is a unifying set of characteristics that reach across the three tropes. I have defined these characteristics as the following: an emphasis on creativity as a fundamental form of knowledge creation; an understanding of creativity as something that is inherently situated and embodied; and finally a recognition of the ethical and moral responsibilities of creativity, the radical intent at the heart of becoming other-wise. With these characteristics comes a wider, deeper shift in ontology that is reflected in the debate around metamodernism. Although I'll leave a more in-depth analysis of metamodernism for later, I think that at its core, the idea that our understanding of reality as a lived experience has undergone a profound change since the new millennium is an important one. While the post-modern period embraced the essential constructedness of existence, ultimately denying any contact with the 'real', this new period has at its heart a yearning for a new kind of ethical and emotive contract with objectivity. The global crises of the Anthropocene and the social and political fallout from the perceived failings of neoliberalism have given new impetus to the prioritisation of the sensorium of lived experience, what Martin Heidegger famously called 'the worlding of the world' (Bolt 2011, 87). The paradigm shift in creativity, from Kearney's metaphor of the postmodern labyrinth, to a new fourth period, one I've provisionally identified as the *map/rhizome/string-figure*, is just part of this much wider cultural transformation. And this transformation does not exclude the sciences. In their introduction to *The Handbook of Science and Technology Studies*, Felt *et al.* recognise that current global issues necessitate a fundamental reassessment of how academic expertise and knowledge engages with the world (2017, 24).

I would argue that this wider contextual understanding helps expose the underlying imperative for research-informed creativity in the arts and humanities today. I'll be saying far more about creativity as research in the next chapter as well as directly engaging with postdigitality; for now, though, it's enough to recognise at least two things: that the map/rhizome/string-figure offers an interrelated set of tropes or 'images of imagining' by which such creativity might be envisioned and theorised; and that the discussions in this chapter have given us a theoretical base from which we can begin our exploration of postdigital storytelling.

Summary

In China Miéville's novel *The City & the City* (2009), the mysterious city of Orciny is as much a textual construct, conjured up by the illegal treatise, *Between Two Cities*, as any physical instantiation. And yet both books, the real *The City & the City*, and the fictitious *Between Two Cities*, can be considered to be tactical interventions, open invitations to see the world, and our part in it, in a completely different, even radical, way. The same is true of the digital work, *These Pages Fall Like Ash* (2013). As it says at the front of the hardcopy book that came with the project, 'the book is a map, everything in it is a path to be walked'. As we've seen, the work is centred around the reader undertaking an affective, transmedial journey, in which both they and their relationship to urban space are transformed.

This chapter has outlined a number of key foundational concepts and propositions that help us understand works such as *These Pages Fall Like*

Ash, works that embrace both digital and non-digital components; are textual and performative; are collaborative both in construction but also wreaderly experience (Landow 1992); and, as a consequence, are fictional yet self-consciously (auto)biographical.

First and foremost, this chapter has established a revised critical definition of creativity within academia today. To do this, I've refashioned two conceptual precepts, the first, Pope's definition of creativity from 2005, and the second, Kearney's schematic of the paradigmatic shifts in Western European artistic representation, summarised as mirror, lamp and labyrinth (1988).

At the heart of this 'refashioning' has been the fundamental recognition that we are now living in a world in which the dominant cultural logic of postmodernism is over. Where we are is still up for debate but I, along with others, such as van den Akker, Gibbons and Vermeulen, and Mary K. Holland, have found critical traction in the term metamodernism. As this chapter has shown, and as I will continue to explore over the course of the book, metamodernism embraces a number of important aspects of contemporary existence. Perhaps most fundamentally it represents a pivot from scepticism and relativism, to new modes of affective meaning and connection, what Holland calls 'an antidote to the destructiveness of [postmodern] ironic detachment' (2013, 123). Everything I have to say about storytelling rests on the emergence of this new ontological paradigm.

From this critical position, I then crafted a revision of Pope's definition of creativity, an enhancement that places emphasis on creativity as a fundamental form of knowledge creation; an understanding of creativity as something that is inherently situated and embodied; and finally a recognition of the ethical and moral responsibilities of creativity, the radical, affective intent of becoming other-wise. I then put forwards three different ways of thinking about how we might conceptualise creativity in this new modality: first, as a form of map-making; second, as a string figure; and third, as a rhizome. The chapter concluded by positing the map/rhizome/string-figure triptych as the fourth category in Kearney's schematic, a means by which creativity within the metamodernist modality may be represented and understood.

Note

1 I'll be discussing Lerner's *10:04* in much more detail in Chapter 6.

Works cited

Abba, T. and Speakman, D. *These Pages Fall Like Ash*. Bristol: Circumstance, 2013.

van den Akker, R. Gibbons, A. and Vermeulen, T., eds. *Metamodernism: Historicity, Affect, and Depth After Postmodernism*. London: Rowman & Littlefield International, 2017.

van den Akker, R. and Vermeulen, T. 'Periodising the 2000s, or, the Emergence of Metamodernism' . *Metamodernism: Historicity, Affect, and Depth After Postmodernism*. Eds, van den Akker, R. Gibbons, A. and Vermeulen, T. London: Rowman & Littlefield International, 2017, 1–19.

Arts and Humanities Research Council. 'Definition of Research'. Web. 5 April 2018. https://ahrc.ukri.org/funding/research/researchfundingguide/introduction/definitionofresearch.

Barthes, R. 'The Death of the Author'. *Aspen* 5–6, 1967.

Barrett, E. 'Introduction'. *Practice as Research: Approaches to Creative Arts Enquiry*. Eds, Barrett, E. and Bolt, B. London: I.B. Tauris, 2010, 1–14.

Barrett, E. and Bolt, B., eds. *Practice as Research: Approaches to Creative Arts Enquiry*. London: I.B. Tauris, 2010.

Baudrillard, J. *Impossible Exchange*. Trans. Turner C. London: Verso, 2001.

Baudrillard, J. *Simulacra and Simulation*. Trans. Glaser, S. Ann Arbor: University of Michigan Press, 1994.

Bohm, D. *On Creativity*. London: Routledge, 1996.

Borges, J. 'On Exactitude in Science'. *Collected Fictions*. Ed, Borges, J. London: Penguin, [1946] (1999), 1.

Bolt, B. *Art Beyond Representation: The Performative Power of the Image*. London: I.B. Tauris, 2004.

Bolt, B. *Heidegger Reframed*. London: I.B. Tauris, 2011.

Bolt, B. 'The Magic is in Handling'. *Practice as Research: Approaches to Creative Arts Enquiry*. Eds, Barrett, E. and Bolt, B. London: I.B. Tauris, 2010, 27–34.

Brewster, A. 'Beachcombing: A Fossicker's Guide to Whiteness and Indigenous Sovereignty'. *Practice-led Research, Research-led Practice in the Creative Arts (Research Methods for the Arts and Humanities)*. Eds, Smith, H. and Dean, R. T. Edinburgh: Edinburgh University Press, 2009, 126–149.

Casone, K. 'The Aesthetics of Failure; Post-Digital Tendencies in Contemporary Computer Music'. *Computer Music Journal*, 24 (4), 2000: 12–18.

Cultural Mapping Toolkit. Partnership of Legacies Now and Creative City Network of Canada. 2010. We. b. 2 March 2018. www.creativecity.ca/database/files/library/cultural_mapping_toolkit.pdf.

D'Ancona, M. *Post-Truth: The New War on Truth and How to Fight Back*. London: Ebury Press, 2017.

Deleuze, G. *Difference and Repetition*, Trans. Patton, P., New York: Columbia University Press, [1968] 1994.

Deleuze, G. and Guattari, F. *A Thousand Plateaus*. Trans. Massumi, B. London: Athlone, [1980] 1988.

Deleuze, G. and Guattari, F. *What is Philosophy?* Trans. Burchell, G. and Tomlinson, H. London: Verso, [1991] 1994.

Dobrin, S. I., ed. *Writing Posthumanism, Posthuman Writing*. Anderson, South Carolina: Parlor Press, 2015.

Facer, K. and Pahl, K. 'Introduction'. *Valuing Interdisciplinary Collaborative Research: Beyond Impact*. Eds, Facer, K. and Pahl, K. Bristol: Polity Press, 2017. 1–21.

Felt, U., Fouché, R., Miller, C. A. and Smith-Doerr, L., eds. *The Handbook of Science and Technology Studies*. Cambridge, Massachusetts: Massachusetts Institute of Technology, 2017.

Felt, U., Fouché, R., Miller, C. A. and Smith-Doerr, L. 'Introduction to the Fourth Edition'. *The Handbook of Science and Technology Studies*. Eds, Felt, U., Fouché, R., Miller, C. A. and Smith-Doerr, L. Cambridge, Massachusetts: Massachusetts Institute of Technology, 2017, 1–26.

Fowles, J. *The French Lieutenant's Woman*. London: Jonathan Cape, 1969.

Gibbons, A. 'Contemporary Autofiction and Metamodern Affect'. *Metamodernism: Historicity, Affect, and Depth After Postmodernism*. Eds, van den Akker, R. Gibbons, A. and Vermeulen, T. London: Rowman & Littlefield International, 2017, 117–130.

Gibbons, A. 'Postmodernism is Dead. What Comes Next?'. *Times Literary Supplement*. 12 June 2017. Web. 8 March 2018. www.the-tls.co.uk/articles/public/postmodernism-dead-comes-next.

Green, B. *The Elegant Universe: Superstrings, Hidden Dimensions and the Quest for the Ultimate Theory*. London: Vintage, 2000.

Haraway, D. J. *Staying with the Trouble: Making Kin in the Chthulucene*. Durham: Duke University Press, 2016.

Harvey, D. *The Condition of Postmodernity*. Oxford: Blackwell Publishing, 1990.

Hawkins, H. *Creativity*. London: Routledge, 2017.

Holland, M. K. *Succeeding Postmodernism: Language and Humanism in Contemporary American Literature*. London: Bloomsbury, 2013.

Jackson, S. *Patchwork Girl; or a Modern Monster by Mary/Shelley and Herself*. Watertown, MA: Eastgate Systems, 1995.

Jameson, F. *Postmodernism or, The Cultural Logic of Late Capitalism*. London: Verso, 1991.

Kearney, R. *On Stories*. London: Routledge, 2002.

Kearney, R. *Poetics of Imagining: Modern to Post-Modern*. Edinburgh: Edinburgh University Press, 1998.

Kearney, R. *The Wake of Imagination*. London: Taylor and Francis, 1988.

Kirby, A. *Digimodernism: How New Technologies Dismantle the Postmodern and Reconfigure Our Culture*. London: Continuum, 2009.

Landow, G. P. *Hypertext: The Convergence of Contemporary Critical Theory and Technology*. Baltimore: John Hopkins University, 1992.

Latour, B. *Down to Earth: Politics in the New Climatic Regime*. Cambridge: Polity Press, 2018.

Lerner, B. *10:04*. London: Granta, 2015.

McClanahan, A. *Dead Pledges: Debt, Crisis, and Twenty-First Century Culture*. Stanford: Stanford University Press, 2017.

Miéville, C. *The City & the City*. London: Macmillan, 2009.

Miéville, C. *London's Overthrow*. London: The Westbourne Press, 2012.

Miéville, C. 'Reports of Certain Events in London'. *Looking for Jake and Other Stories*. London: Macmillan, 2005, 53–75.

Moraru, C. *Cosmodernism: American Narrative, Late Globalization and the New Cultural Imaginary*. Ann Arbor, Michigan: The University of Michigan Press, 2011.

Moulthrop, S. 'Rhizome and Resistance: Hypertext and the Dreams of a New Culture'. *Hyper/Text/Theory*. Ed, Landow, G. P. Baltimore: John Hopkins University Press, 1994, 299–319.

Paul, C. and Levy, M. 'Genealogies of the New Aesthetic'. *Postdigital Aesthetics: Art, Computation and Design*. Eds, Berry, D. M. and Dieter, M. London: Palgrave Macmillan, 2015, 27–43.

Peck, J. 'The Cult of Urban Creativity'. *Leviathan Undone? Towards a Political Economy of Scale*. Eds, Keil, R. and Mahon, R. Vancouver: University of British Columbia Press, 2009, 159–176.

Pope, R. *Creativity: Theory, History, Practice*. London: Routledge, 2005.

Ryan, M., Foote, K. and Azaryahu, M. *Narrating Space / Spatializing Narrative: Where Narrative Theory and Geography Meet*. Columbus: Ohio State University Press, 2016.

Salami, M. *John Fowles' Fiction and the Poetics of Postmodernism*. Vancouver: Fairleigh Dickinson University Press, 1992.

Smith, H. and Dean, R. T. 'Introduction: Practice-led Research, Practice-led Practice – Towards the Iterative Cyclic Web'. *Practice-led Research, Research-led Practice in the Creative Arts (Research Methods for the Arts and Humanities)*. Eds, Smith, H. and Dean, R. T. Edinburgh: Edinburgh University Press, 2009, 1–38.

Smyth, K., Power, A. and Martin, R. 'Culturally Mapping Legacies of Collaborative Heritage Projects'. *Valuing Interdisciplinary Collaborative Research: Beyond Impact*. Eds, Facer, K. and Pahl, K. Bristol: Polity Press, 2017, 191–213.

Steiner, G. *Grammars of Creation*. London: Faber and Faber, 2001.

Tabbi, J. 'Introduction'. *The Bloomsbury Handbook of Electronic Literature*. Ed, Tabbi, J. London: Bloomsbury, 2018, 1–9.

Tringham, R. 'Creating Narratives of the Past as Recombinant Histories'. *Subjects and Narratives in Archaeology*. Eds, van Dyke, R. M. and Bernbeck, R. Boulder, Colorado: University Press of Colorado, 2015, 27–54.

Tuer, D. *Mining the Media Archive: Essays on Art, Technology and Cultural Resistance*. Toronto: Y Y Z Books, 2006.

Turchi, P. *Maps of the Imagination: The Writer as Cartographer*. San Antonio, TX: Trinity University Press, 2004.

3 Postdigital storytelling

Introduction

In the previous chapter, I positioned discussions of contemporary storytelling within a wider, and deeper, debate about creativity. Rather than understanding creativity as a monolithic given, I showed how the ontology that underpins artistic representation has gone through at least three paradigmatic shifts, what Kearney identifies as mirror, lamp and labyrinth (1988). His third category, the labyrinth of postmodernism, is a key period in the development of electronic literature (Hayles 2008). Yet, as I also indicated, it is now accepted that the era of postmodernity is over. What this new period should actually be called is still open to debate; I, along with others (van den Akker, Gibbons and Vermeulen; Holland), adopt the term metamodernism. Much of the previous chapter was about understanding what this new cultural paradigm portended in regards to creativity. Critically, I offered the map/rhizome/string-figure as the new creative modality of the metamodern, the fourth paradigm in Kearney's schematic.

This chapter continues and enriches that analysis in two important ways. First, it positions postdigitality as a fundamental characteristic of this new, metamodernist sensibility and by doing so broadens existing, technologically-determined, definitions of the postdigital. Second, the chapter explores the impact of postdigitality on the epistemological assumptions underpinning academic research, particularly within the arts and humanities. It foregrounds the interrelationship between the map/rhizome/string-figure as a creative modality, and the epistemological exigencies of transdisciplinarity. It ends with a conceptual framework in which postdigitality is deconstructed as a materialising practice within a practice-praxis-research continuum.

Before I turn to any of these things, however, let's start our journey with a shed. Two sheds in fact.

Metamodernism as postdigital: a tale of two sheds

Figure 3.1 is a graphical depiction of the internet in 2015. One way of looking at it is as an idealised representation of postmodern cyberspace – detached,

Figure 3.1 The internet 2015, graphical map.
Source: © Barrett Lyon/The Opte Project. CC BY-NC 4.0. Used with permission.

fragmented, virtual (Harvey 1990). Figure 3.2 shows an artistic representation of this postmodern perspective in the form of Cornelia Parker's art installation, *Cold Dark Matter: An Exploded View*. Constructed in 1991, two years after the birth of the world wide web, the installation consists of the fragments of a garden shed exploded by the British Army. It is atomised, caught midway in its own destruction. Both the loss of the original object and its dynamic parsing into some other mediatised form are key aspects of the work.

But let's take another look at Figure 3.1. Surely another way of interpreting what we see is as a representation of the emergent ontological structuring of metamodernism. In this case what we are witnessing is not the atomisation, fragmentation and indeterminacy of postmodernism as represented by Parker's exploded shed, but rather the map/rhizome/string-figure of situated connectedness, a boundless, transformational ecology of human experience. In this new cultural modality, out goes Parker's *Cold Dark Matter*, and in comes something like Simon Starling's *Shedboatshed (Mobile Architecture No. 2)* (2005), shown in Figure 3.3.

Figure 3.2 Cold Dark Matter: An Exploded View by Cornelia Parker (1991), installation.

Source: © Tate, London 2019. Courtesy the artist and Frith Street Gallery, London.

Shedboatshed has a complexity that is belied by its form. To make the artwork, the artist dismantled an old shed on the banks of the River Rhine and turned it into a boat. Loaded up with the remains of the shed, the boat was then paddled down the Rhine to a museum in Basel. There it was dismantled and re-made back into the shed again, now bearing the scars of its journey.

Figure 3.3 Simon Starling, *Shedboatshed (Mobile Architecture No. 2)*, 2005.
Source: © Simon Starling. Installation view at Museum für Gegenwartskunst, Basel, 2005. Photo Martin P. Bühler.

Made fourteen years after Parker's work, *Shedboatshed* makes no attempt to deconstruct what the artwork is attempting to show. Instead, the shed becomes a kind of artefactual story, a physical inscription of its own interconnectedness with the landscape through which it passed and with those who made it and brought it to the gallery. A word that is often used when describing *Shedboatshed* is pilgrimage. And of course a pilgrimage is a journey story, a connection of people, objects and places, a self-conscious act of transgression in becoming other-wise.

These two sheds – one defiantly iconoclastic, the other, a 'reaching out' or connection with the situated embodiment of the world – neatly encapsulate the shift in creative modality from labyrinth to map/rhizome/string-figure that is at the heart of this book, a transformation that is also captured by the multivalence of Figure 3.1 as labyrinth or metamodernist entanglement.

Yet however we decide to interpret this startling map of the internet, perhaps the most striking thing remains what it actually shows: the profound and complex nature of our interaction with technology. The depth and scale of this interdependency, of course, has increased rapidly since the advent of modern industrialisation in the nineteenth century. Indeed, it is impossible to understand concepts such as modernism or postmodernism without seeing them as fundamental responses to overwhelming technological change (Danius 2002; Jameson 1991). I'm going to be saying more about this in the following chapter when I examine early hypertextual stories within the context of postmodernity.

If we are witnessing the birth of a new, emergent ontology, then Figure 3.1 suggests that something equally profound is happening in terms of our relationship to technology. van den Akker and Vermeulen (2017, 15), in their analysis of metamodernism, recognise the rise of mobile computing and social media as transformational developments, both socially and culturally, but also in the development of market-based capitalism. The overall tenor of their discussion of technology remains sceptical though, drawing on well-established postmodern concerns with ideological control and what Baudrillard described as 'the disappearance of the real' (1994, 145).

Yet if this is indeed a time of a profound readjustment in our relationship to technology, as the notion of metamodernism suggests, then it seems to me that we need an alternative to the sort of postmodern scepticism offered by van den Akker and Vermeulen. I would argue that we must recognise the role that digital technology can play in creating affective, transformative experiences – in other words, narratives that respond to the creative modality of map/rhizome/string-figure. The concept of postdigitality allows us to do that.

One of the first to use the term was the composer, Kim Casone. In an article written at the beginning of this century, Casone argued that a new era was emerging, one in which 'the revolutionary period of the digital information age has surely passed' (2000, 12). At the heart of Casone's thesis was an artistic disenchantment with digital technology, a sense that a new creative period had arrived in which artists were no longer willing to simply accept the perceived superiority of digital form and expression. Casone observed that one outcome of this was the increasing willingness of musicians to add distortion and glitches into their recordings, deliberately re-emphasising a productive and technological process that digital technology naturally erased from the listening experience. For bands such as *Massive Attack* and *Portishead*, for example, this even amounted to replicating the distortion of non-digital technology, in this case the analogue scratch of a needle moving over vinyl. Florian Cramer calls this the rise of DIY ethics and maker culture and sees it as a critical part of a 'post-digital' age (2018, 367). Indeed, for both Casone and Cramer, the avant-garde and the radical is now associated with, what Cramer calls, a 'post-digital hacker attitude' (2015, 20), in which the distinction between old and new media is broken down and remediated (see Nathan Jurgenson's critique of 'Digital Dualism' 2011).

These changes in the early years of this new century were, at least in part, driven by the recognition that digital technology was increasingly ubiquitous, overwhelming every aspect of our lives. As we've seen, the release of the first iPhone in 2007 is a key marker here, particularly in terms of the speed and impact of these developments. The phenomenal success of smartphone and tablet computing has provided a global network the like of which has never been seen or experienced before. This sense of the *normalisation* of digitisation is a key part of the postdigital, to the point where, as Cramer says, '"digital" has become a meaningless attribute' (2018, 362–363). We are postdigital therefore, clearly not because we have moved on from being digital in any sense, but because of the exact opposite condition – its ubiquity has made any need for explicitness redundant. As David M. Berry and Michael Dieter note of this postdigital condition, '[c]omputation becomes experiential, spatial and materialized in its implementation, embedded within the environment and embodied, part of the texture of life itself but also upon and even within the body' (2015, 3). Just as pervasive connectivity has made redundant the distinction between 'offline' and 'online', then so too has the word 'digital' become outdated, conjuring up 'a disjuncture in our experience that makes less and less sense in the experience of the everyday' (Berry and Dieter 2015, 2–3). For both Cramer, and Berry and Dieter, postdigital refers to the condition where this binary divide between digitality and non-digitality is no longer representative of how we actually experience our lives. Instead, postdigital offers a new kind of epistemological mash-up, what Berry and Dieter call 'a hybridised approach towards the digital and the non-digital, finding characteristics of one within the other' (2015, 6).

Postdigitality is therefore both a condition of ubiquity as well as technological affordance. This gives postdigital an inherent tension between hybridity on the one hand and the technological pervasivity on which such hybridity rests on the other: or, to put it more crudely, between hegemonic computationalism versus situated, embodied empowerment. Both Casone's and Kramer's definition of postdigital captures this tension, this 'resistance against'/'coming together'. Postdigital hybridity is therefore as much a reaction against digitality as it is an embrace of new structures of possibility.

According to Mel Alexenberg, while the digital age prioritised the two-fingers of the qwerty keyboard or mouse, the postdigital period invites 'the rediscovery of ten fingers' (2011, 35). The postdigital is therefore tactile and physical as much as computational, inhabiting the creative spaces between the digital, biological, cultural, and spiritual. For Alexenberg, this sense of entanglement predominates, a messy human-computational interaction which not only emerges from our postdigital condition, but also strives to represent and challenge it in new ways.

While these definitions of postdigitality offer an effective baseline, I would suggest that they remain rather limited. What I hope this book has started to put forward in these opening chapters is a far stronger conception of postdigitality, a phenomenon grounded in the rise of a new dominant ontology. Indeed, it is the contention of this book that postdigitality remains one of metamodernism's most striking features.

Berry and Dieter are right of course that such entanglement is certainly a product of technology, its ability to support a mobile, cross-platform hybridity, where the user is no longer rooted to a static PC somewhere in an office. It's equally clear that Kramer's postdigital hacker attitude has engendered a cultural receptivity to hybridity as a creative act. Yet, I would argue that it is the emphasis on the body as the locus of affective experience that is absolutely critical here; postdigitality fundamentally encapsulates the changing relationship between technology and human behaviour within the map/rhizome/string-figure paradigm. Unlike notions of posthumanism, for instance, which stress the absolute and unfaltering imbrication of human existence and digital technology (Dobrin 2015; Hayles 1999), I would argue that the postdigital offers a more nuanced approach, in which human creativity still retains the capability to move actively and self-consciously across the digital and non-digital domains. Critical to posthumanism is what Dobrin calls the 'moment of inquiry in which the human subject is called into question via its imbrications with technologies such as cybernetics, informatics, artificial intelligence, genetic manipulation, psychotropic and other pharmaceuticals, and other bio-technologies' (2015, 3). The posthuman invites a postliterary future, one where the artist 'discovers that no content is now immediately identified with his innermost consciousness' (Agamben 1999, 54). It is this calling into question of the human subject that I see as the divide between posthumanism and the postdigital. The postdigital does not seek to call into question the human subject as a meaningful locus of understanding. Rather, like the concept of metamodernism itself, it strives to find a way of re-establishing human agency, of a re-engagement with the imperative of our existence through, what Heidegger termed, 'worlding the world' (Bolt 2011, 87). As Berry notes:

> In the midst of a world which has become blurred and ungraspable, the postdigital constellation becomes a primary element, an object for a cultural analytics that provides connection and a sense of cohesion in a fragmentary digital experience.
>
> (2015, 51)

It is the postdigital which conforms to and enriches our understanding of creativity in the age of the map/rhizome/string-figure. After all, string figures are far-more effective when you have ten fingers to play with. Postdigitality, then, rather like Golumbia's concept of 'computationalism', stands in some kind of opposition to the posthuman. Yet, while 'computationalism' ultimately denies our capability to use and adapt technology to mankind's greater good, seeing it as tethering us to the ongoing production of 'problematic psychological and/or ideological formations' (Golumbia 2009, 225), postdigital offers at least the possibility of real, meaningful and beneficial change. As Caroline Bassett notes, the focus of the postdigital is less about technology *per se* than about 'newly materialized worlds' (2015, 137).

Central to this concept of the postdigital, then, is a belief in the possibility of human agency, in the sort of resistance that Alexander Galloway saw possible in his analysis of internet protocols and infrastructure (2004). Although Galloway recognised the internet 'as the most highly controlled mass media hitherto known' (147), he did not deny that a certain level of resistance was possible, in particular what he termed hacking, tactical media (including computer viruses) and internet art. In his more recent study of the French philosopher François Laruelle, Galloway notes that artistic and creative freedom can only operate when 'intimately allied with history and with the real exigencies of matter' (2014, 81).

It is in the new dawn of metamodernism that artists work with these 'real exigencies of matter', developing new hybridities of poetics (analogue/digital) and praxis (interdisciplinary/transdisciplinary). Freedom therefore comes not through the technofetichism of posthumanism with its dismantling of the liberal humanist subject, as Hayles would have it (1999, 246), but rather with a poetics and praxis predicated on the new creative paradigm of map/rhizome/string-figure, a dichotomy captured in the transition from the rupturing metamorphosis of Parker's *Cold Dark Matter* (Figure 3.2) to the embodied latencies of Starling's *Shedboatshed* (Figure 3.3).

Postdigital storytelling: some definitions

And so, finally, to the crux of this work: postdigital storytelling. Drawing on my revised conception of postdigitality outlined in the opening section of this chapter, I would propose the following five points in terms of how the postdigital can be defined and understood:

1. that the postdigital is a 'coming together', a hybridisation of both the digital and the non-digital domains, and a denial of any implicit 'disjuncture', to use Berry and Dieter's term, in how we experience them;
2. that this 'coming together' or hybridisation has two vectors: the movement of the non-digital to the digital and the digital to the non-digital. In other words, postdigital is not simply the recognition of digital ubiquity. Rather, it is an understanding of something more subtle: that at the heart of the postdigital lies a more open and fluid negotiation between the digital and the non-digital;
3. in this revised configuration of postdigitality, the postdigital operates from two states or positions: *within* or *across* the digital/non-digital nexus;
4. that regardless of state or position (within or across the domains), the postdigital remains the dominant modality;
5. that the non-digital domain is subordinate to the digital domain.

These five points provide the basis of postdigital that I'll be using in this book. Point two is especially important in that it recognises postdigitality as a spectrum running from digitality to non-digitality, with movement, or negotiation, coming from both ends. A study of postdigitality could therefore equally come through an analysis of non-digitality, investigating the impact of ubiquitous digitality on the material, non-digital, world. Point three expands on this by stating that postdigitality is therefore something that operates in a number of states: within or across. The first, *within*, recognises that the postdigital can be seen to operate within the two binary nodes of the nexus, in other words, within the digital domain or within the non-digital domain, but not across the two. The 'within' captures this domain specificity. The second state, *across*, represents a mode of postdigitality where domain specificity no longer applies. Here, postdigitality is understood as working across the two nodes or domains, in accordance with either vector outlined in point two (digital to non-digital, non-digital to digital). In other words, postdigital is explicated in the material hybridisation of digital and non-digital elements, what Steve Benford and Gabriella Giannachi call 'spatial tapestries' in their analysis of mixed reality performance (2011, 69).

Postdigital *storytelling* should therefore be understood as operating in these two states, both *within* and *across* the digital/non-digital nexus. However, what I'm also going to be arguing is that our binary model of digital/non-digital domains is not without hierarchy. In fact, these domains exist in, what I term, a *dominant/subordinate* relationship as given in point five. In other words, hybridity is not an equal playing field, a simple case of mixing and matching across a transmedial spectrum. Until very recently, it was the non-digital domain that was dominant: here, that which was digital was only perceived from within this non-digital domain. In other words, digital stories were perceived as 'not' printed texts, their very form defined by deficit. However, this book argues that, crucially, this hierarchy is now in the process of being reversed; indeed, that it is this reversal of the *dominant/subordinate* relationship that is a key aspect of postdigitality. Now, it is the digital domain that is dominant and it is from this position that the non-digital is subordinately considered. In our postdigital age, then, it is the printed book that has become a signifier of functional deficit: in other words, printed text have come to represent 'that which is not digital' rather than any natural or intrinsic form of published work in and of itself. Crucially, too, this hierarchical modelling is primarily an engendering of social-cultural phenomena rather than simply a reflection of technological ubiquity. Just as any recourse to digital storytelling was necessarily a strategic decision by both writer and reader in the first hierarchical mode, so any recourse to print-based media in the second mode has become equally strategic and transgressive. This transformation of hierarchical mode forms an important context to what I have to say about postdigital poetics over the coming chapters, and in particular the artistic striving for, what Danius has called, 'authentic experience' (2002, 3).

My primary focus in this book is the impact of postdigitality on digital storytelling. In other words, I intend to position my study primarily from the perspective of *digital storytelling*, looking at how postdigitality can be understood from *within* the digital domain (in other words, how postdigitality is encoded within digital works) and *between* the domains (the degree to which digital storytelling embraces non-digital elements). Over the coming chapters I will be investigating hypertextual fictions, locative mobile storytelling, and collaborative and social-media based narratives, in other words, the sort of 'data-led, locative, generative, algorithmic, sensor-based … personalised, proximal, augmented, real-time, time-sensitive, adaptive, collaborative, and share-y' stories that *Editions at Play* extoll. Yet, in order to present as fuller analysis as possible, I'll also be examining the impact of postdigitality on non-digital forms, too, the sort of thing that Alison Gibbons undertakes, for example, in her analysis of experimental fiction in her book, *Multimodality, Cognition, and Experimental Literature* (2012). I've already introduced Ben Lerner, a writer who has lots to say about hybridisation and transmediality, not only through his two novels, but also his more experimental printed works such as *Blossom* (2015), a collaboration with the artist Thomas Demand. Over the course of this book the printed (non-digital) works of other writers, such as Zadie Smith, Joanna Walsh and Mark Z. Danielewski will also be discussed in regard to the impact of postdigital hybridisation on the traditionally-printed novel. In this way, my analysis seeks to ask some fundamental questions about the impact of this digital/non-digital nexus, including, 'What is digital storytelling in the postdigital era?' and, 'What impact does hybridisation and transmediality have on our understanding of digital storytelling?' Through these and other questions, our understanding of how and why storytelling more generally is being transformed will provide important insights into the formation and characteristics of postdigitality and, through that, to a clearer and deeper conceptualisation of metamodernism itself.

One valid starting point, then, in terms of my definitions, is the concept of *digital storytelling*. If the focus of this work is primarily digital storytelling in the postdigital condition then I need a clear understanding of what digital storytelling has traditionally meant before I can begin to look at how and why this definition has been transformed since the beginning of this century.

In its broadest interpretation, digital storytelling is simply what the name implies: the telling of stories through digital technologies. This could include dedicated web-based stories, kindle books, blogs (factual and non-factual), tweets, video and gaming, embracing both amateur and professionally-produced content, standalone, software dependent and web-based openly accessible (see Alexander 2011, 3).

Yet what I've already addressed in the previous chapter indicates that if my analysis is to have real critical depth then it needs to be firmly situated within wider discussions of creativity. It also must recognise that this analysis is specific in that it is focussed on the use of postdigital storytelling as a form of transformative research within the arts and humanities, drawing on what I've already said about storytelling as research in the previous chapter. Clearly, these all act as important delimiters to any notion of digital storytelling as a conceptual definition. Let's remind ourselves of how these considerations were brought together in my revised explication of creativity:

> Creativity is a situated and embodied act, in which, through the (co-) creation of new knowledge, perceptions and understanding of the world are changed.

For the purposes of this book, then, digital storytelling as a concept has the following baseline characteristics:

1. It is embraces an awareness of the situated and embodied nature of creativity.
2. It leads to change through knowledge creation.

Within these two characteristics are some of the key aspects of the map/rhizome/string-figure modality: a situated interconnectedness; a drift away from postmodern scepticism towards a more nuanced ontological position driven by the ethical imperatives of responding to increasingly complex global crises. This is a new kind of storytelling, a storytelling for the postdigital, metamodernist condition of the twenty-first century.

At one level, taking this critical perspective does not seem to alter very much. Yet this should not fool us into believing that nothing has changed. Instead, the transformation has come through in what isn't quite so visible, namely the underlying purpose behind the construction of any digital story, just as we saw with *Shedboatshed*. In other words, it's not the 'what' that will necessarily see the impact of these ideas, but rather the 'why'.

The term digital storytelling is useful here because of its emphasis on the *process* of storytelling, in other words how stories are actually put together, disseminated and experienced. This triptych recalls Paul Ricoeur's circle of triple mimesis by which he attempts to illuminate the way in which narrative represents and influences human action. Ricoeur argued that narrative representation rests on three stages of interpretation: mimesis1 (prefiguration of the field of action), mimesis2 (configuration of the field of action) and mimesis3 (refiguration of the field of action). Mimesis1 describes the prefiguring of our life-world as it seeks to be told; mimesis2 concerns the creative configuring of the text; while mimesis3 refers to the refiguring of our actual, lived existence after encountering the text (see Kearney 2002, 133). Like Ricoeur's circle of triple mimesis, the term digital storytelling embraces both the creative act itself (the 'digital' making) but also its telling, in other words, how the story is experienced by the reader. It is this telling, what Ricoeur would describe as the refiguring of our existence (mimesis3), that aligns digital storytelling with the sort of knowledge construction, the becoming other-wise, that is so critical to our map/rhizome/string-figure ontology. I would therefore argue that digital storytelling is not a weak term, lost to generality, but its exact opposite: an overtly strategic one with particular resonance today.

Other descriptors have not been quite so lucky in their longevity: 'electronic literature' is one of the less fortunate. As a term, electronic literature has a respected pedigree, of course. The *Electronic Literature Organisation* (ELO) was formed in 1999 as a means of supporting both the scholarship and the cataloguing of digital fiction. Its *Electronic Literature Collection* (Vols 1–3) provides an indispensable archive of digital fiction from the late 1990s while its annual conference is an important event in the digital fiction calendar. Yet

I would argue that both the term itself and its underlying philosophy (certainly as applied by organisations such as the ELO) have been superseded, undercut by newer, more innovative approaches.

In her analysis of the *Electronic Literature Collection*, N. Katherine Hayles notes that when the ELO set up a committee to define electronic literature as a term, they came up with the following: 'work with an important literary aspect that takes advantage of the capabilities and contexts provided by the stand-alone or networked computer' (2008, 3). The term 'important literary aspect' reveals a lot here. The Oxford English Dictionary defines 'literature' as 'written works, especially those considered of superior or lasting artistic merit', and this emphasis on some notion of artistic quality is mirrored in the ELO's own definition. I would suggest that this prioritisation of literature (rather than fiction, hypertext or storytelling) preserves the concerns of digital authors from the 1990s about the status and artistic credibility of their work. Despite attempts at being inclusive and open, there is a sense of an elitist agenda at work through both the ELO and the wider discourse of electronic literature. This certainly comes across from Hayles' analysis. In *Electronic Literature: New Horizons for the Literary*, Hayles oscillates between the ELO's own, more tightly defined, definition and broader conceptions of her own. Hayles' defines electronic literature as 'a first-generation digital object created on a computer and (usually) meant to be read on a computer' and is a significant opening-up of the ELO definition (2008, 3). As Hayles goes on to say:

> Hypertext fiction, network fiction, interactive fiction, locative narratives, installation pieces, 'codework,' generative art, and the Flash poem are by no means an exhaustive inventory of the forms of electronic literature, but they are sufficient to illustrate the diversity of the field, the complex relations that emerge between print and electronic literature, and the wide spectrum of aesthetic strategies that digital literature employs.
>
> (2008, 30)

Hayles offers a generous updating of ELO's position; yet at its heart it maintains a focus on the digital artefact as literary text. Research that follows on from this position is very much engaged with the digital text itself through the application of critical theory. In other words, electronic literature tends towards an inward-looking solipsism, following in the traditions of literary and compositional studies more generally (see Koehler 2017).

This is not to say that electronic literature is not without radical intent. As we'll see in the next chapter, the sort of hypertext fictions that form the backstay of the *Electronic Literature Collection* were certainly experimental in design and purpose. Indeed, the poet and academic, Loss Pequeño Glazier (2001), in his analysis of digital poetics, is not alone when he argues that electronic literature is best thought of as a continuation of experimental print literature, pushing at the boundaries of innovation.

A good example of this is *10:01* by Lance Olsen and Tim Guthrie, available through the *Electronic Literature Collection*, volume one.[1] The story was first published as a hard copy novel in 2005 before being adapted into a hypertext by Tim Guthrie in the same year (2005). The story is based around a random collection of thirty-seven people sitting in a cinema watching the trailers (see Figure 3.4). The trailers run for ten minutes and one second before the start of the film. The story is experimental in its unstructured design, forcing the reader to create a coherent narrative by moving from person to person, or/and by using the timeline. As Mehdy Sedaghat Payam says in her analysis of the story, *10:01* self-consciously deploys present tense, numbered sentences and film-script format: 'each one is a transgression from established narrative practice, drawing our attention toward the language itself and how few of the available modalities have been, and are being, used by traditional novels' (2018, 315).

The story is innovative in the sense that it uses the affordance offered by hypertext to create a non-chronological multimedial narrative. The story is very much about the inherent limitations of language and narrative structure. It is avant-garde and postmodern, a digital celebration of the underlying constructiveness of experience. In a story such as *10:01* we are still very much in the labyrinth.

Figure 3.4 Screenshot from *10:01* by Lance Olsen and Tim Guthrie (2005).
Source: © Lance Olsen and Tim Guthrie.

Electronic literature, then, can certainly be innovative. However, this notion of radicalism is very much grounded in ideas and concepts of literary experimentalism, in the way the digital artefact enhances or adapts existing understanding through what might be considered to be the literary avant-garde. This is not a problem in itself, of course. Digital work with this sort of agenda has and will continue to produce exciting and important work, as this book will demonstrate. Yet, as I've argued in the previous chapter, this is certainly not the whole story. My discussion of creativity, and the wider ontological changes associated with the map/rhizome/string-figure schematic, has meant that terms such as electronic literature, alongside hypertext and interactive fiction, now seem rather old fashioned, both in the nature of the works they represent but also the implicit assumptions behind their design.

As we'll see, storifying platforms such Twine and Genarrator have led to the increasing democratisation of digital storytelling, circumventing the traditional gatekeepers of electronic literature. In their discussion of Twine, for example, Astrid Ensslin and Lyle Skains adapt George Landow's term 'wreadership' (1992), a neologism that describes the active role of both the writer and the reader in the production of the text, and retool it for a new age.

> [Twine's] accessibility and facility have led to its rise as a vehicle for personal exploration of narrative experiences in underserved populations, establishing hypertext outside of the experimental art realm and into the mainstream.
>
> (2018, 303)

Ensslin and Skains not only emphasise the transformation brought about by programmes such as Twine and even Storyspace, but also participatory social media platforms, creatively empowering both the individual and their communities (2018, 304). A good example of this is Teju Cole's small fates project (2011). Drawing on the tradition of *faits-divers*, Cole began to tweet a series of obscure stories drawn from local newspapers about ordinary people living in Nigeria, all within the limit of 140 characters: 'Mrs Ojo, of Akure, favoring a cashless economy, bought goats, beans, and onions in Sokoto, and paid with bags of marijuana' (19 August 2011). I'll be saying a lot more about this, and other social-media projects, in Chapter 7. However, it's worth saying that for Cole, it was exactly the fragmented and elliptical ordinariness of these tales that he felt was best expressed through Twitter. In other words, the aesthetic and dynamic affordance of the platform (including the enmeshment of retweets, replies and likes) was a key aspect of the storytelling process.

I would argue that if there is to be a poetics of the postdigital condition, then it must embrace this sort of bottom-up, democratisation of the storytelling process, rather than simply limit itself to the formalised experimentalism of electronic literature. In other words, it must address the imperatives raised in my discussion of map/rhizome/string-figure. Collaborative, co-produced works are paramount here, recognising storytelling as a critical intervention in the formation of individual and communal identity. As Kearney notes, 'We are made by stories before we ever get round to making our own … That is why narrative is an open-ended invitation to ethical and poetic responsiveness' (2002, 154). It is this becoming other-wise, this ethical responsiveness, that lies at the heart of postdigital storytelling.

It's worth saying that there is a long and respected tradition of collaborative digital storytelling associated with the *Center for Digital Storytelling* at the University of California, Berkeley (Lambert 2013), something we'll be looking at in more detail in Chapter 7. Here digital storytelling refers to the way digital technology can be used to construct and disseminate personal (auto)biography (Dunford and Jenkins 2017). My concept of postdigital storytelling includes (auto)biography, but it also goes much further, at the very least embracing the more radical hybridisation of fact and fiction that we've already met through Gibbons' discussion of autofiction ('Contemporary Autofiction' 2017) and Brewster's personally-situated writing. Whereas the *Center for Digital Storytelling* is very much founded on what Nancy Thumim identifies as the twin discourses of democratisation and therapy, postdigitality broadens and deepens any understanding of digital storytelling, reformulating the term as a key attribute of the postdigital, post-postmodern, condition. As we'll see, postdigital storytelling as used here therefore refers to a much wider set of activity, embracing a multitude of platforms and technical affordances, while at the same time reaching across a panoply of narrative form and genre.

Yet, whatever definition you take – digital storytelling or electronic literature – recourse to the digital demands a different way of thinking about text and narrative. For example, here's Katherine N. Hayles on one of the key differences between traditional print-based and digital texts:

> Whereas all the words and images in the print text are immediately accessible to view, the linked [digital] words … become visible to the user only when they appear through the cursor's action. Code always has some layers that remain invisible and inaccessible to most users. From this we arrive at an obvious but nevertheless central maxim: print is flat, code is deep.
>
> (2004, 75)

'Print is flat, code is deep' is a powerful way of summarising the issue at stake here. In other words, unlike traditional print, digital storytelling always consists of invisible layers of code, hidden from the surface aesthetics of what you actually see on the screen. A similar model is represented in David M. Berry's 'digital iceberg' (2015, 47).

The most obvious example of this would be hypertext markup language (HTML) which, alongside cascading style sheets (CSS) and JavaScript, forms the hidden computer language of the world wide web. App developers use a range of programming languages, from Python and Java, to languages such as C++. Yet beneath all these coding languages for apps, the web and other standalone programmes, lies the binary 0s and 1s of machine code and the alternating voltages associated with it. Any computer-mediated text is therefore layered in a way that the traditional text is not, consisting of text (code) that users are never expected to see or directly engage with (Hayles 2008, 163). While some authors of digital stories might be able to programme using HTML, Python or Java themselves, more complex projects often demand a team-based approach, in which the author works alongside creative and digital developers. This was the case for Joanna Walsh's *Seed* (2017), for example, a project that entailed collaboration between the author, illustrator (Charlotte Hicks) and a large design and development team at *Visual Editions* (Hawlin 2017).[2] It's a similar situation for Kate Pullinger's *Breathe* (2018).[3] For less complex projects, other options are available. Many digital authors reply on development platforms that negate any reliance on programming. For example, *Adobe Dreamweaver* is a powerful web development tool, while there are a range of proprietary platforms that support apps development. The storifying platforms Twine, Genarrator and Storyspace fall into this category too. And if digital stories are written on blogging and social media platforms (such as Twitter and Facebook) then no additional programming knowledge is needed at all of course, they simply co-exist in what has become the normalised space of everyday digital interaction.

What I've outlined here are strategies by which authors can distance themselves from the need to programme, should they wish too. Of course, these different approaches do not change the fact that all computer-mediated text is fundamentally a product of computer code. Hayles' notion of layering has not gone away simply because we're accessing a story through Facebook or an app on our smartphone. For Francisco J. Ricardo this therefore raises the central issue of what actually constitutes digital art: is it the aesthetic *experience* of what is seen and experienced, or is it the aesthetic *mechanism* of the art itself (2009, 6). Ricardo calls these two positions 'the algorithmic experience and the moment of aesthetic enactment' (2009, 6). Clearly, any digital artform incorporates both an underlying algorithmic mechanism (the programmable software) and the surface aesthetic experience. The more complex a digital artform, the more dependent it becomes on customised coding and design, as we've seen with *Seed* and *Breathe*. The approach taken in this book is the sort of pluralism adopted by Ricardo himself, in other words, a critique that moves from the poetic/descriptive to the functional/mechanistic, as appropriate. Drawing on our revised definition of postdigital creativity, this book therefore builds on other work in its development of a coherent critical 'ontology' for postdigital storytelling, and through that, creativity more broadly. Yet, although underlying software and coding is important, our primary focus remains both the *poetic* affordances and the *praxical* and *research* application of postdigital storytelling within the (post) digital humanities.

I would suggest that the two two-layer model identified by Hayles (2008) and Ricardo represents a conceptualisation of digital text from a postmodern, labyrinthine, modality. Here, we're in the world of avant-garde electronic literature such as *10:01*. In order to revise this modal so that it fits with the map/rhizome/string-figure modality, we need to add a layer of *materiality*, in other words the conceptual space through which a digital artefact becomes postdigital. As I've already argued, one of the key features of postdigitality is what I term a 'coming together', or the hybridisation, of both the digital and the non-digital (point one in my definition). It's in this *layer of materiality* that such 'coming together', or entanglement, occurs. This could include physical non-digital forms (hardcopy print, for example); but it could just as well refer to physicality, in which the digital becomes part of an (augmented) physical performance. *Empire Soldiers VR* by the theatre company Metro Boulot Dodo, described in the Introduction, is a good example of this, as is *These Pages Fall Like Ash* (2013). The key takeaway point, however, is that in the postdigital condition, a digital object is conceptualised as being inextricably intertwined with the non-digital world. Both *Empire Soldiers VR* and *These Pages Fall Like Ash* are in effect transmedial performances where the digital component of each is exactly that: just one part of a larger whole in which the digital and non-digital parts are in constant flux and negotiation.

There's one further layer that has slipped through much of the analysis to date and that's *data*. A lot of digital platforms naturally capture and record significant amounts of data about the user. This is especially the case for social media platforms, of course. Facebook stores login locations and device details, as well as a host of other online activities. Website cookies allow companies to track and record online behaviour. The use of digital platforms to record and monitor online activity is so embedded in the computationalism (to use Golumbia's term) of our postdigital world that it is now impossible to disaggregate data capture from the functionality of many digital systems. Yet it does seem as though public toleration of this behaviour has reached some kind of threshold. Revelations concerning the harvesting of the private data of 80 million people through Facebook by the data analytics firm, *Cambridge Analytica*, in 2018 has emboldened both public and government outrage (Hindman 2018), as has growing fears of homophilous sorting and a retreat into post-truth echo chambers (D'Ancona 2017, 49). Some kind of instrumental change in how such personalised data is collected and used seems likely, building on existing legislation, such as the European Union's General Data Protection Regulation (GDPR).

In fact there are hints that things are changing already. Web-based platforms such as Twine and Storyspace do not record any user data; more generally, tentative proposals are being put forward in which user-generated data is regarded as the private property of those who generated it (*The Economist* 2018). For most digital storytelling, of course, data creation is an inadvertent side effect of particular host platforms or software, rather than being a primary purpose of the story. Yet this is not always the case. *Sea Hero Quest* (2016) is a digital story set up explicitly to capture individual (anonymised) data in order to contribute to research on dementia. The app is downloaded onto a mobile device. The user completes the game as well as inputting a basic set of personal data, including age, gender and ethnicity. The user is made aware about what data is captured, how it is anonymised and stored, and to what purpose it will be put. To date, the original app has had over three million users. A virtual reality (VR) version is now also available. Of course, this explicit generation and use of data will not apply to most digital stories. Yet stories like *Sea Hero Quest* offer a glimpse of a possible future, a future in which the exchange of personal data is done openly and only at the behest of the user in return for identifiable research outcomes.

In summary, this section has set out a few key markers in terms of how digital storytelling might be understood and theorised in the postdigital condition. I've made two arguments in regard to this, the first concerning *breadth*, and the second, *depth*. The first, *breadth*, refers to the need to extend existing definitions of digital narratives, breaking down barriers and divisions. Established terms such as electronic literature and hypertext fiction are indelibly associated with postmodern avant-garde experimentalism. Although significant in their own right, both terms have had their day, steeped as they are in the imperatives of the previous century. In the creative paradigm of map/rhizome/string-figure, a new definition is needed to embrace a postdigital world. For now, until something better comes along, postdigital storytelling will do. As expounded here, postdigital storytelling embraces both the situated and embodied nature of creativity, as well as its primary function of generating change, the becoming other-wise already highlighted. Postdigital storytelling, then, defines a much wider set of digital activity than previous terms such as electronic literature. This activity includes a wide range of platforms and software, poetry and prose, fiction and non-fiction, text-based and multimodal. It incorporates both the production of the digital artefact, but also its wider 'wreaderly' affect.

If map/rhizome/string-figure heralds a new kind of breadth to traditional notions of digital storytelling, then it also elicits a modified understanding of depth. By depth I refer to the notion of layering identified by Hayles (2008). As we've seen, Hayles reduces the technical architecture of any digital artefact to a series of textual layers – from the aesthetic surface of the screen (Ricardo's aesthetic engagement), to the hidden code of the software. What I've proposed here is an extension of Hayles' formulation, deepening and enriching the layering in two ways. The first is to recognise that in the reformulation of postdigitality, the implicit disjuncture between the digital and non-digital domains no longer applies. Instead, we have what I have called a 'coming together' between the two, a condition that has led to new kinds of transmedial hybridity. This is a new layer of *materiality*, a conceptual space that recognises the interplay between digital and non-digital elements and by which conceptions of both are transformed.[4] Mel Alexenberg calls this the return of ten fingers, but it's actually much wider than that, embracing all aspects of embodied physicality. The idea of materiality *returning* to us, as opposed to being *new*, is actually much closer to the truth. A layer of materiality always existed with digital objects; it's just that its narrative potential has now been legitimated both by the map/rhizome/string-figure modality and by changing technological affordance.

As we'll see, mobile technology and pervasive connectivity have transformed how stories are experienced. Even more recent developments in the area of VR and augmented reality have accelerated this phenomenon, transforming the nature of human interaction with digital technology. I use the term performance here both to refer to this technologically-mediated interplay between narrative and space (what Ryan, Foote and Azaryahu call narrating space/spatialising narrative) and to the kind of mixed-media artistic performances described by Benford and Giannachi. Both tropes are a recognition of the radical new ways that performative storytelling is pushing the boundaries in terms of the situated and embodied nature of narrative.

The second addition to Hayles' formulation is to recognise the intimate connection between mobile, pervasively connected software and the generation of personalised data. Digital stories may not self-consciously set out to create and use such data (as *Sea Hero Quest* does, for example); yet for those digital authors who use proprietary platforms such as social media applications or smartphone apps, the reality is that any user interaction will create some kind of record of their activity.

These two additions give us the revised four-layered model for postdigital storytelling shown in Figure 3.5.

Each layer can only interact directly with the layer or layers to which it is adjacent. While the code and data remain hidden, the narrative (aesthetic enactment) and materiality layers are entangled to varying degrees with the real (human) world. As has been explained, the focus of this book will be the two outer layers of our model: I have argued that it is here, in narrative (textual) design and embodied materiality, that the poetic affordance of postdigital stories lies. Yet clearly, any critical approach needs to be cognizant of the two hidden layers of data and code; certainly in this book I'll be undertaking a consideration of software where appropriate. As Ricardo notes, this is a pluralist position, a position that both advances the notion of underlying universalist characteristics and principles behind digital storytelling, but one that also acknowledges the need to recognise the individual specificities of software (code/data) functionality.

Postdigital storytelling as research

Understanding the poetics of postdigital storytelling is clearly important. However, a key, and at times, controversial issue still remains to be discussed, namely the relationship between postdigital storytelling and research. To

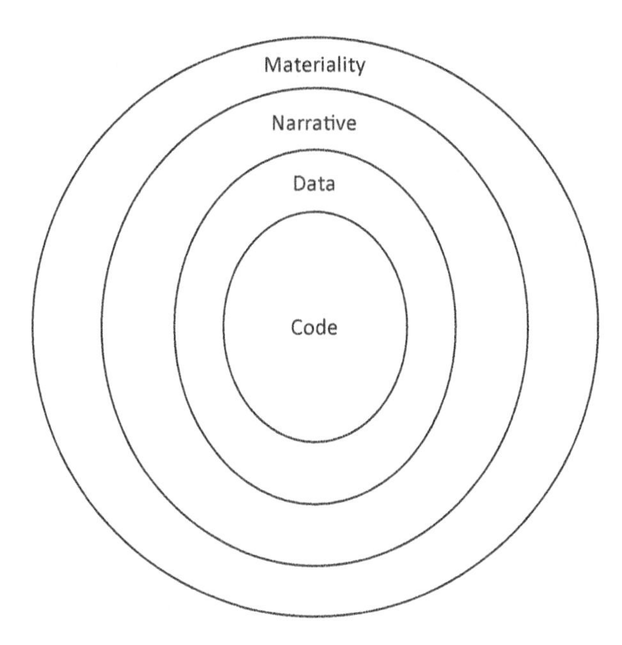

Figure 3.5 Revised four-layer model for postdigital storytelling.
Source: © Author.

put it more bluntly, how and to what extent can postdigital storytelling be considered a form of research within academia. Although there has been a lot of discussion concerning creative writing as research (for example, Webb 2015; Dawson 2005), digital storytelling of any kind is noticeably absent from these debates. Where it is discussed the focus is rather narrow; in the case of Koehler, for example, the analysis is limited to exploring the relationship between composition, creative writing studies and the digital humanities.

This is a significant lacuna that this book seeks to fill. More specifically, I argue here that the paradigm shift of artistic representation associated with map/rhizome/string-figure must necessitate a much broader and systemic reformulating of practice-based research and methodology. Although traditionally seen as an individual activity, the sort of research projects outlined by Facer and Pahl point to practice-based approaches that have collaboration and interdisciplinarity at their heart. Yet I would argue that even these projects do not fully embrace the radical challenge and potential of metamodernism. To do that we need to move beyond the interdisciplinary and multidisciplinary boundaries that we have set ourselves, towards that which is between, across and beyond. As Ruth Finnegan says, '[t]he view of research as essentially a creature of the universities is notably a partial one' (2005, 5). In other words, we need to become transdisciplinary in our approach (Nicolescu 2008).

But first things first. Before I start to consider the specifics of postdigital storytelling as research, I need to pull back our focus to where I began this chapter, in other words to a consideration of creativity more generally.

> Creativity is a situated and embodied act, in which, through the (co-) creation of new knowledge, perceptions and understanding of the world are changed.

Although I've spent some time reformulating a new definition of creativity, what I have not discussed is creativity as academic research. If creativity is indeed to be considered a form of research then it must, in some way, engender new knowledge that can be shared and generalised beyond the individual or individuals undertaking the activity. In other words, the research must lead to some kind of change in wider understanding and impact, wider in the sense of reaching beyond those involved in the research activity itself. This is especially so for academic research, as opposed to the sort of privately-funded research that takes place in organisations and companies such as Google and Tesla. In publicly-funded universities, the imperative is that research is outward facing, openly accessible and shareable, with tangible beneficial impact on society.

When a creative artefact is made, of course, that making will always produce new knowledge generated by the practice. This is praxical knowledge, knowledge generated through doing, and it is at the heart of all creative practice. Yet it is praxical knowledge that is limited to the creator(s). The same applies to those viewing, reading or otherwise engaging with the artefact as 'audience'. Such interaction with any creative artefact could (and perhaps should) lead to some kind of new knowledge or understanding, however small; but once again, it is personalised and hidden, perhaps not even aware of itself. For praxis to lead to research, then, praxical knowledge of process needs to be explicated in some way that builds on from the individualised specifics of its making or/and engagement. In other words, it needs to fit the criteria for academic research that I've set out above: that it leads to the generation of new knowledge, that this knowledge is generalisable, and of wider public interest and impact. We'll see how this can actually happen further on in this chapter. However, the key message at this point is a simple one, and one worth repeating: creativity *per se* is not research (although as we'll see, it can be part of, or the basis of, a research project).

I hope this is not a controversial statement. While Candy differentiates between practice-based and practice-led research, both involve some kind of critical exegesis that exists separately from the creative artefact itself. It's this exegesis that reveals the process involved in the making. The same is true of Smith and Dean's practice-led and research-led practice (2009, 5). Both taxonomies offer a spectrum of criticality within creative research. In practice-based/practice-led research, for example, the creative artefact forms a substantial part of the output, but it must also include, what Candy calls, 'a substantial contextualisation of the creative work' (2006, 3). With practice-led/

research-led on the other hand, the creative artefact is used to inform a much more substantial critical exegesis, and it is this exegesis which forms the key output. Of course, these divisions are always artificial in one way or another, as Smith and Dean recognise; indeed, their formulation of what they call an 'iterative cyclic web' is just one attempt at clarifying the underlying synergy between the two approaches (2009, 19–25).

Our understanding of creative practice is very much indebted to Martin Heidegger's concept of 'praxical knowledge' (Bolt 2011) and it is these ideas and approaches that inform the core of my own approach to understanding the relationship between creative practice and research. Indeed, Bolt states that Heidegger's thinking provides nothing less than 'an artist's guide to the world' (2011, 12). For Heidegger, human consciousness and understanding (Dasein) are only achieved through our physical and emotional engagement with the world, a 'being-in-the-world', rather than through objective theorisation (2011, 3). In other words, 'the world is discovered through Dasein's involvement or handling of it. This constitutes what Heidegger terms the "worlding of the world"' (2011, 87). This is a term we've already met before, informing Moraru's cosmodernism and, through that, van den Akker, Gibbons and Vermeulen's concept of metamodernism. According to Heidegger, while our 'being-in-the-world' generally consists of mundane chores and activities, art provides a space in which to think about ourselves more deeply. Central here is an ethical imperative, the ability to make a difference to everyday lives (2011, 127).

It is this emphasis on practice over theory that forms the foundation of *praxis* as a concept. 'Praxical knowledge involves a reflexive knowing that imbricates and follows on from handling' (Bolt 2010, 34). Bolt uses the term 'materialising practices' to describe this kind of knowledge creation (Bolt 2004). Such 'materialising practices' are embedded not only within the creation of the artefact itself, but also in its performative consumption. Bolt is clear that practice must involve 'theoretical cognition if it is not to remain blind' (86). In other words, praxis does not simply refer to artistic creation in and of itself; crucially, it also involves theoretical engagement as part of its materialising practice. However, such praxical knowledge gained through the materialising practice of any artwork is still not research. There is still no sense here of the sort of academic research I've described, leading to the production of new knowledge, that is generalisable, and of wider public interest. For it to be so, there must be some kind of critical exegesis, as I've already argued. A creative artefact may incorporate all sorts of innovatory techniques and approaches; yet, as Bolt notes, fundamentally, 'it is not the job of the artwork to articulate these, no matter how articulate that artwork may be. Rather, the exegesis provides a vehicle through which the work of art can find a discursive form' (2010, 33). The critical exegesis, then, is central to praxis-informed research, providing an alternative form by which new and emergent knowledge can be brought to bear on existing theories and ideas (Bolt 2011, 156).

In a modification to Smith and Dean (2009) and Bolt (2010, 2011), I suggest that any creative activity can be broken down into a natural progression from practice, to praxis, to research where,

1. **Practice** refers to the technical skill or craft involved in making;
2. **Praxis** refers to the reflexive application of theory to the practice, forming praxical knowledge and
3. **Research** refers to the wider theorisation of the praxical knowledge.

Both practice and praxis can lead to the generation of new kinds of knowledge. This is knowledge that is formed through doing. However, at levels one and two of my progression, such experiential knowledge is private, individuated knowledge. It is not academic research as it stands (based on my definition). For this to happen there needs to be written some kind of critical exegesis by which the private realm of personal experience is turned outwards, discursively engaging with established theory (level three).

We need look no further than the UK's Arts and Humanities Research Council (AHRC) for a confirmation of these ideas. In their definition of research, they state that 'research activities should primarily be concerned with research *processes*, rather than *outputs*' (AHRC, my italics). By outputs the AHRC is referring to artefactual output, the specifics of things or stuff, as it were, rather than the transferable knowledge involved in their making or production. The AHRC go on to state:

> Creative output can be produced, or practice undertaken, as an integral part of a research process as defined above. The Council would expect, however, this practice to be accompanied by some form of documentation of the research process, as well as some form of *textual analysis or explanation* to support its position and as a record of your critical reflection. *Equally, creativity or practice may involve no such process at all, in which case it would be ineligible for funding* from the Arts and Humanities Research Council.
>
> (AHRC, my italics)

The AHRC are clear: a project whose only outcome is a creative artefact will not be funded. The Australian Research Council (ARC), funded through the Australian Government, takes a similar position. Process rather than output. To help with this demarcation, the AHRC specify three key areas any research funded by them must address: (1) it must define a series of research questions, issues or problems; (2) it must specify a research context for these questions and (3) it must specify the research methods for addressing and answering the research questions. The AHRC's position does not negate the role of storytelling as both output and methodology: as we've already seen, storifying is one of the key approaches highlighted by Facer and Pahl in their analysis of the Connected Communities Programme (2017). Yet the AHRC stress that the creative output must inform, and be informed by, a set of wider questions and research imperatives.

This focus on 'process' chimes strongly with Bolt's notion of materialising practices (2004). Drawing on Heidegger's ideas of 'handling' and 'concernful dealings', Bolt outlines an ontology of artistic practice that foregrounds its performative, as opposed to representational, nature (186). At the centre of this performativity is the idea that the practice of art is always embodied, both in its making and reception. Through this material (embodied, situated) process 'the outside world enters the work and the work casts its effects back into the world' (190). For Bolt, the work of art is therefore this embodied process of change, rather than simply the representational artefact itself: 'it is through this dynamic and productive relation that *art emerges as a revealing*. The work of art is this movement' (188, my italics).

This concept of creativity as a process of revealing, a materialising practice, aligns strongly with what I've been arguing in terms of creative practice. Perhaps just as importantly, it reaffirms my position in terms of arts-based research: if creativity is a materialising practice, then it is here, in these processes of change and revealing, Heidegger's 'worlding of the world', that any research needs to be grounded.

As we've seen, postmodernity, with its emphasis on the constructed nature of truth, pragmatism and irony, denies our ability to see beyond the fragmentary; here, as Harvey notes, 'coherent representation and action are either repressive or illusionary' (1990, 52). Yet what I've argued in this chapter is that we have entered a new cultural paradigm, one I've termed metamodernism. Associated with metamodernism is the creative modality of the map/rhizome/string-figure. Within this new modality, then, the notion of emotional and ethical 'affect' is given renewed legitimacy, empowering situated and embodied-based concepts such as Bolt's materialising practice and Heidegger's 'worlding'. Within metamodernism, creative practice is freed from postmodern 'incredulity or irony' and instead grounded in new found understandings of 'faith, conviction, immersion and emotional connection' (Konstantinou 2017, 93). In this new modality, artistic practice is no longer simply representational but rather *tactical* or *instrumental*, 'designed to do something to us' (Konstantinou 2017, 94). Metamodernism therefore legitimates the ontological position where artistic affect and technical craft are materially intertwined. As such, it is here, in the analysis of such affective intervention and wider materialising practices, that any (practice-based) research needs to focus.

What we can see, then, is that any new creative paradigm legitimates new research methods and methodologies. Under postmodernism, research was very much informed by a hermeneutical scepticism by which texts were understood to hold no meaning in and of themselves. Here, research into practice was an exploration of depthlessness with its denial of deeper symbolic authenticity. The text as an authentic, affecting agent did not, could not, exist. As such, the focus of any exegesis was inward looking and solipsistic, drawing on the established methods and methodologies of literary criticism. In essence, the artist constructed the artefact and then deconstructed it as 'text' in the critical exegesis.

Conversely, as we've seen, the metamodernism of map/rhizome/string-figure reintroduces creative practice as the locus of affective action. The artefact is no longer treated with suspicion *a priori,* and in consequence the role of the exegesis is changed too, freed from the reductionist imperative to tease out, what Rita Felski calls, a text's 'hidden causes, determining conditions, and noxious motives' (2015, 12). Rather than positioning ourselves behind the text 'we might place ourselves in front of the text, reflecting on what it unfurls, calls forth, makes possible' (2015, 12). The implications for practice-based research, and in particular the critical exegesis, are profound.

Adjusting our focus back to the particularity of postdigital storytelling does not alter any of these conclusions. In his analysis of creative writing in the 'new humanities', Dawson strongly criticises what he sees as the traditional focus on writing technique and publishing alongside exegeses that simply summarise the creative process.

> It is not by reasserting authorial intention as the basis of critical evaluation that writers will claim intellectual authority within the academy, but by exploring the political and discursive effects of their literary products and accepting responsibility for them.
>
> (2005, 214)

In some ways, Dawson's fears in regards to creative writing are being addressed. What should already be clear from this book is the degree to which creative writing/storytelling/storifying has become recognised as a fundamental aspect of knowledge creation. Both this, and the movement away from postmodern irony, depthlessness and suspicion, have had the effect of pushing storytelling in new and exciting directions.

One of the most important developments here is the inclusion of digital and non-digital storytelling as part of postdigital hybridisation which has opened up new opportunities for *transdisciplinary* research (Nicolescu 2008). Whereas *multidisciplinarity* refers to the studying of a research topic from several disciplines concurrently, and *interdisciplinarity* to the studying of a research topic involving the active transfer of methods from one discipline to another, transdisciplinarity 'concerns that which is at once between the disciplines, across the different disciplines, and beyond all disciplines' (Nicolescu 2008, 2). For Alfonso Montuori, transdisciplinarity is characterised by a focus on the practical applications of knowledge in which the enquirer is recognised as an active and ethical participant:

> [T]he method of transdisciplinarity is "in vivo" [within the living]: the knower is not a bystander looking at knowledge in its pristine cognitive state, but an active participant, a *being-in-the-world.* The

> Transdisciplinary approach does not focus exclusively on Knowing, but on the inter-relationship between Knowing, Doing, Being and Relating.
> (2008, xi)

It is transdisciplinarity's alignment with the concept of materialising practices that makes it such a valuable contribution to our discussion of research approaches within the map/rhizome/string-figure modality. Although Facer and Pahl do not use the term, preferring instead the more recognised label interdisciplinarity, the ethos of transdisciplinarity hovers over many of the research projects funded through the AHRC's Connected Communities Programme.

Facer and Pahl offer four broad tropes of collaborative, cross-disciplinary research that have emerged through the programme: (1) *mutual learning*, in which research output occurs through the embodied learning of the participants; (2) *crowd and open*, built on large scale (community-based) data collection and collective work; (3) *design and innovation*, utilising publics/communities/users in the design of services, policies or (technical) products and (4) *correcting the record*, addressing social/cultural inequalities and silences. Although not presented as exhaustive in any way, these four tropes speak of emergent forms and structures that are transforming research practice, methods and methodologies across the spectrum of interdisciplinary/transdisciplinary approaches. Equally importantly they offer an articulation of the wider (collaborative) role that (postdigital) storytelling can play within interdisciplinary/transdisciplinary projects.

The growing complexity of the challenges facing humankind suggest that transdisciplinary research will only increase in importance. When faced with global crises such as those associated with the Anthropocene, for example, including climate change, and social and cultural resilience, we need to be able to think outside of established discipline boundaries. As Montuori states, 'disciplinary fragmentation is not just a *response* to knowledge, it actually *frames* knowledge' (2008, xv). In trying to deal with complexity head on, transdisciplinarity seeks to overcome such fragmentation by looking both *across* and *beyond* academe, re-establishing the primacy of an ethically-grounded and embodied understanding of knowledge construction, *in vivo*, within the living.

In the reconfiguration of map/rhizome/string-figure, storytelling, whether digital or non-digital, operates within an enhanced ontological landscape. This landscape includes both the singular creative project and exegesis, such as Brewster's personally-situated writing project (2009), but also more complex inter/transdisciplinary projects. In this latter mode, storytelling is just one component of a much wider set of research activities. For example, the AHRC-funded project we've already looked at, *Preserving Place: A Cultural Mapping Exercise*, involved both academics from a range of disciplines (including history, literature, geography, computing and health sciences) and non-academic representatives from various community and heritage groups

(Smyth *et al.* 2017). Cultural mapping included qualitative and quantitative methods, of which interviewing and storifying, using both digital and non-digital forms, were but one part. Research output consisted of the mapping toolkit rather than any formal critical exegesis.

What seems clear then is that in the new creative modality of map/rhizome/string-figure, the established division between creativity and exegesis becomes much more permeable than otherwise envisaged, certainly by Dawson, but also Barrett and Bolt, and Smith and Dean. At the heart of the traditional exegesis is the sort of critical apparatus that Felski, in her book *The Limits of Critique* (2015), systematically deconstructs until it stands before us as nothing more than 'a repertoire of stories, smiles, tropes, verbal gambits, and rhetorical ploys' (7). For Felski, critical theory, with its hermeneutic suspicions and interrogations, is but one way of thinking about a text, one that is neither a privileged method nor with any greater claim to truth or insight. The real danger of critical theory is that it comes with the assumption that whatever is not critical theory must therefore be *uncritical*. This Felski can't allow. In her search for a post-critical consensus, Felski makes an impassioned plea for a new way of understanding creative practice:

> [W]orks of art cannot help being social, sociable, connected, worldly, immanent – and yet they can also be felt, without contradiction, to be incandescent, extraordinary, sublime, utterly special. Their singularity and their sociability are interconnected, not opposed.
>
> (2015, 11)

Any critical exegesis both needs to reflect this way of thinking about creative artwork *per se*, as well as engaging with the fact that interdisciplinarity and transdisciplinarity by their very nature demand new and innovative ways of approaching complexity. Transdisciplinarity, in particular, places special emphasis on reaching beyond entrenched knowledge structures and taxonomies. In this regard, the traditional exegesis might not be appropriate to satisfy the research outputs of a transdisciplinary project; rather, the critical component could be re-fashioned, for example, transformed into some new fusion of creative and critical elements, such as the cultural mapping toolkit (see Hilevaara and Orley 2018 on creative/critical writing). Stephen Benson and Clare Connors, in their overview of creative criticism, state how '[c]reative criticism, in short, is writing which seeks to do justice to what can happen – does happen; will happen; might or might not happen – when we are with an artwork. We can call that being-with an encounter …' (2014, 5). For them, such 'encountering' leads to change and creative criticism is the 'writing out' (5) of this process. The artwork is an embodied encounter, then, a materialising, revealing or practice, to use Bolt's term (2004), which requires a more adaptive response to the creative/critical divide than has been the norm. As Benson and Connors state, '[c]reative criticism does not propose to do away with … this demarcation [between the creative and critical] but it does seek variously to test its workings and its reach, and to imagine innovative forms for the taking of sides' (2014, 24).

I would argue that this testing of workings and reach, the visualisation of new and emergent forms of research, are energised in the modality of map/rhizome/string-figure. In our condition of postdigital normativity, digital creativity has particular resonance within any transdisciplinary endeavour. As I've already described, postdigital storytelling cannot be reduced to a consideration of the sort of formal experimentalism associated with electronic literature. Postdigital storytelling is much wider, embracing both formal and informal modes and forms, such as social media and open web-based platforms; it embraces digital biography and non-fiction as much as fiction, prose and poetry. It is code, data, narrative and performance. Yet, as transmedia projects like *These Pages Fall Like Ash* (2013) prove, the postdigital is also what is not digital. In this new form of mash up, unexplored potentialities open up, intermingling the digital and the non-digital, the performative and the textual, creating new artistic domains. As we've seen, such transformations of creative modality demand equally new and innovative forms of research methodology by which output and impact can be articulated and maximised. It is to the addressing of these issues of postdigital poetics and praxis that this book will turn. Yet before we do so, it is necessary to understand the foundations on which our current creative modality is built, a period in which digital storytelling was born.

Summary

This chapter has established postdigitality as one of the key characteristics of our current time. van den Akker, Gibbons and Vermeulen provide a powerful introduction to metamodernism, yet their discussion of digital technology is limited, drawing on the sort of cynical misgivings of Golumbia's 'computationalism'. While this has some value, of course, I have argued here that a more nuanced approach is not only possible, but also powerfully re-legitimated within the ontological boundaries of map/rhizome/string-figure. Crucially, I have proposed that *postdigital* normativity offers a new conceptual layer to the metamodern. By doing this I have necessarily deepened and extended accepted notions of postdigitality. While Casone, Berry and Dieter and Cramer (2018) offer a technological-determinist perspective, I have tried to understand the term as a key aspect of a new ontology. As part of this new creative modality, I have understood postdigitality as something that is both *within* and *across* the digital/non-digital nexus. In that sense, it's as much of a condition of any single domain (a print-only text, for example, or a digital-only story) as it is of works that are self-consciously hybridic and transdomainal. If we are to comprehend postdigitality in its entirety then we need to be conscious of what is happening *within* domains as much as what is happening *across*.

From this perspective, then, postdigitality is just not about technological advances. Instead, what has emerged from this chapter is a concept of postdigitality as a key marker of deep, sustained social and cultural change.

Bruno Latour notes that, '[w]hat is certain is that we can no longer tell ourselves the same old stories' (2018, 44). Understanding postdigitality as a critical aspect of new and emergent forms of storytelling remains central here. As Gibbons in her chapter on contemporary autofiction notes, '[i]n a crisis-ridden world, subjects are once more driven by a desire for attachment to others and to their surroundings' (2017, 130). In the face of global catastrophe, Haraway's 'making trouble' and Pope's 'becoming other-wise' give 'worlding the world' a new kind of political imperative, impossible in the ironic depthlessness of postmodernity. In the modality of map/rhizome/string-figure, creativity becomes a tactical intervention whose end goal is affective change. Postdigitality offers a conceptual framework by which these interventions can be theorised and understood.

Finally, the chapter explored the implication of these conclusions on the way postdigital storytelling can be used within academic research. Building on Heidegger's concept of 'worlding the world' (Bolt 2011), the chapter explored the interrelationship between practice, praxis and research. Two outcomes were critical here. First, that any research must involve some kind of critical exegesis in which creativity as a materialising practice, inducing change and revealing, is explored. Second, that interaction along the practice-praxis-research axis takes place within a transdisciplinary context, in which boundary and domain transgression (*across* and *beyond*) is legitimated. Such transformations of postdigital creativity will not only engender new forms of storifying but also radical forms of creative-critical output.

Yet before we start to explore this new terrain, it is important that we first think more deeply about postdigital storytelling itself as a form of narrative making. Postdigital storytelling did not simply spring into life sometime in the new millennium; instead, it has a long and involved history, both within and across our digital/non-digital domains, as should already be clear. Understanding that history is crucial to how we might think more deeply about our own postdigital world and the role and function of creativity within it.

Notes

1 See http://collection.eliterature.org/1/works/olsen_guthrie__10_01/1001.html.
2 See https://seed-story.com.
3 See https://editionsatplay.withgoogle.com/#!/detail/free-breathe.
4 See my discussion of 'embodied space' in Chapter 6 in regards to locative mobile storytelling.

Works cited

Abba, T. and Speakman, D. *These Pages Fall Like Ash*. Bristol: Circumstance, 2013.
Agamben, G. *The Man without Content*. Stanford: Stanford University Press, 1999.

van den Akker, R. Gibbons, A. and Vermeulen, T., eds. *Metamodernism: Historicity, Affect, and Depth After Postmodernism*. London: Rowman & Littlefield International, 2017.

van den Akker, R. and Vermeulen, T. 'Periodising the 2000s, or, the Emergence of Metamodernism'. *Metamodernism: Historicity, Affect, and Depth After Postmodernism*. Eds, van den Akker, R. Gibbons, A. and Vermeulen, T. London: Rowman & Littlefield International, 2017, 1–19.

Alexander, B. *The New Digital Storytelling: Creating Narratives with New Media*. Oxford: Praeger, 2011.

Alexenberg, M. *The Future of Art in a Postdigital Age: From Hellenistic to Hebraic Consciousness*. Bristol: Intellect, 2011.

Arts and Humanities Research Council. 'Definition of Research'. Web. 5 April 2018. https://ahrc.ukri.org/funding/research/researchfundingguide/introduction/definitionofresearch.

Barthes, R. 'The Death of the Author'. *Aspen* 5–6, 1967.

Barrett, E. 'Introduction'. *Practice as Research: Approaches to Creative Arts Enquiry*. Eds, Barrett, E. and Bolt, B. London: I.B. Tauris, 2010, 1–14.

Barrett, E. and Bolt, B., eds. *Practice as Research: Approaches to Creative Arts Enquiry*. London: I.B. Tauris, 2010.

Bassett, C. 'Not Now? Feminism, Technology, Postdigital'. *Postdigital Aesthetics: Art, Computation And Design*. Eds, Berry, D. M. and Dieter, M. London: Palgrave Macmillan, 2015, 136–150.

Baudrillard, J. *Impossible Exchange*. Trans. Turner C. London: Verso, 2001.

Baudrillard, J. *Simulacra and Simulation*. Trans. Glaser, S. Ann Arbor: University of Michigan Press, 1994.

Benford, S. and Giannachi, G. *Performing Mixed Reality*. Cambridge, Massachusetts: MIT Press, 2011.

Benson, S. and Connors, C. 'Introduction'. *Creative Criticism: An Anthology and Guide*. Eds, Benson, S. and Connors, C. Edinburgh: Edinburgh University Press, 2014, 1–47.

Berry, D. M. 'The Postdigital Constellation'. *Postdigital Aesthetics: Art, Computation And Design*. Eds, Berry, D. M. and Dieter, M. London: Palgrave Macmillan, 2015, 44–57.

Berry, D. M. and Dieter, M. 'Thinking Postdigital Aesthetics: Art, Computation and Design'. *Postdigital Aesthetics: Art, Computation and Design*. Eds, Berry, D. M. and Dieter, M. London: Palgrave Macmillan, 2015, 1–11.

Bohm, D. *On Creativity*. London: Routledge, 1996.

Borges, J. 'On Exactitude in Science'. *Collected Fictions*. Ed, Borges, J. London: Penguin, [1946] (1999), 1.

Bolt, B. *Art Beyond Representation: The Performative Power of the Image*. London: I.B. Tauris, 2004.

Bolt, B. *Heidegger Reframed*. London: I.B. Tauris, 2011.

Bolt, B. 'The Magic is in Handling'. *Practice as Research: Approaches to Creative Arts Enquiry*. Eds, Barrett, E. and Bolt, B. London: I.B. Tauris, 2010, 27–34.

Brewster, A. 'Beachcombing: A Fossicker's Guide to Whiteness and Indigenous Sovereignty'. *Practice-led Research, Research-led Practice in the Creative Arts (Research Methods for the Arts and Humanities)*. Eds, Smith, H. and Dean, R. T. Edinburgh: Edinburgh University Press, 2009, 126–149.

Candy, L. *Practice-based Research: A Guide*. Sydney: Creativity & Cognition Studios, University of Technology, Sydney, 2006. Web. 4 April 2018. www.creativityandcognition.com/resources/PBR%20Guide-1.1-2006.pdf.

Casone, K. 'The Aesthetics of Failure; Post-Digital Tendencies in Contemporary Computer Music'. *Computer Music Journal*, 24 (4), 2000: 12–18.

Cole, T. 'Small Fates'. March 2011. Web. 21 November 2018. www.tejucole.com/other-words/small-fates.

Cramer, F. 'Post-Digital Writing'. *The Bloomsbury Handbook of Electronic Literature*. Ed, Tabbi, J. London: Bloomsbury, 2018, 361–369.

Cramer, F. 'What is "Post-Digital"?'. *Postdigital Aesthetics: Art, Computation and Design*. Eds, Berry, D. M. and Dieter, M. London: Palgrave Macmillan, 2015, 12–26.

Cultural Mapping Toolkit. Partnership of Legacies Now and Creative City Network of Canada. 2010. Web. 2 March 2018. www.creativecity.ca/database/files/library/cultural_mapping_toolkit.pdf.

D'Ancona, M. *Post-Truth: The New War on Truth and How to Fight Back*. London: Ebury Press, 2017.

Danius, S. *The Senses of Modernism: Technology, Perception, and Aesthetics*. New York: Cornell University Press, 2002.

Dawson, P. *Creative Writing and the New Humanities*. London: Routledge, 2005.

Deleuze, G. *Difference and Repetition*, Trans. Patton, P., New York: Columbia University Press, [1968] 1994.

Deleuze, G. and Guattari, F. *A Thousand Plateaus*. Trans. Massumi, B. London: Athlone, [1980] 1988.

Deleuze, G. and Guattari, F. *What is Philosophy?* Trans. Burchell, G. and Tomlinson, H. London: Verso, [1991] 1994.

Demand, T. and Lerner, B. *Blossom*. London: Mack Books, 2015.

Dobrin, S. I., ed. *Writing Posthumanism, Posthuman Writing*. Anderson, South Carolina: Parlor Press, 2015.

Dunford, M. and Jenkins, T., eds. *Digital Storytelling: Form and Content*. London: Palgrave Macmillan, 2017.

The Economist. 'Should Internet Firms Pay for the Data Users Currently Give Away?' *The Economist*, 11 January (2018). Web. 2 April 2018. www.economist.com/news/finance-and-economics/21734390-and-new-paper-proposes-should-data-providers-unionise-should-internet.

Editions at Play. Web. 5 February 2018. https://editionsatplay.withgoogle.com/#/about.

Ensslin, A. and Skains, L. 'Hypertext: Storyspace to Twine'. *The Bloomsbury Handbook of Electronic Literature*. Ed, Tabbi, J. London: Bloomsbury, 2018, 295–309.

Facer, K. and Pahl, K. 'Introduction'. *Valuing Interdisciplinary Collaborative Research: Beyond Impact*. Eds, Facer, K. and Pahl, K. Bristol: Polity Press, 2017, 1–21.

Felski, R. *The Limits of Critique*. Chicago: University of Chicago Press, 2015.

Finnegan, R. 'Introduction: Looking Beyond the Walls'. *Participating in the Knowledge Society: Researchers Beyond the University Walls*. Ed, Finnegan, R. London: Palgrave Macmillan, 2005, 1–19.

Fowles, J. *The French Lieutenant's Woman*. London: Jonathan Cape, 1969.

Galloway, A. R. *Laruelle: Against the Digital*. Minneapolis: University of Minneapolis Press, 2014.

Galloway, A. R. *Protocol: How Control Exists after Decentralization*. Cambridge, MA: The MIT Press, 2004.

Gibbons, A. 'Contemporary Autofiction and Metamodern Affect'. *Metamodernism: Historicity, Affect, and Depth After Postmodernism*. Eds, van den Akker, R. Gibbons, A. and Vermeulen, T. London: Rowman & Littlefield International, 2017, 117–130.

Gibbons, A. *Multimodality, Cognition, and Experimental Literature*. London: Routledge, 2012.

Gibbons, A. 'Postmodernism is Dead. What Comes Next?'. *Times Literary Supplement*. June 12 2017. Web. 8 March 2018. www.the-tls.co.uk/articles/public/postmodernism-dead-comes-next.

Golumbia, D. *The Cultural Logic of Computation*. Cambridge, Mass.: Harvard University Press, 2009.

Glazier, L. P. *Digital Poetics: Hypertext, Visual Kinetic Text and Writing in Programmable Media*. Tuscaloosa: University of Alabama Press, 2001.

Green, B. *The Elegant Universe: Superstrings, Hidden Dimensions and the Quest for the Ultimate Theory*. London: Vintage, 2000.

Haraway, D. J. *Staying with the Trouble: Making Kin in the Chthulucene*. Durham: Duke University Press, 2016.

Harvey, D. *The Condition of Postmodernity*. Oxford: Blackwell Publishing, 1990.

Hawlin, T. 'No One Gets Out Alive: An Interview with Joanna Walsh'. *Review 31*. 2017. Web. 1 February 2018. http://review31.co.uk/interview/view/19/no-one-gets-out-alive-an-interview-with-joanna-walsh.

Hayles, N. K. *Electronic Literature: New Horizons for the Literary*. Notre Dame, Indiana: University of Notre Dame Press, 2008.

Hayles, N. K. *How We Became Posthuman: Virtual Bodies in Cybernetics, Literature, and Informatics*. Chicago: The University of Chicago Press, 1999.

Hayles, N. K. 'Print is Flat, Code is Deep: The Importance of Media-Specific Analysis'. *Poetics Today*, 25 (1), 2004: 67–90.

Hawkins, H. *Creativity*. London: Routledge, 2017.

Hilevaara, K. and Orley, E., eds. *The Creative Critic Writing as/about Practice*. London: Routledge, 2018.

Hindman, M. 'How Cambridge Analytica's Facebook Targeting Model Really Worked – According to the Person Who Built It'. *The Conversation*, 30 March 2018. Web. 2 April 2018. https://theconversation.com/how-cambridge-analyticas-facebook-targeting-model-really-worked-according-to-the-person-who-built-it-94078.

Holland, M. K. *Succeeding Postmodernism: Language and Humanism in Contemporary American Literature*. London: Bloomsbury, 2013.

Jackson, S. *Patchwork Girl; or a Modern Monster by Mary/Shelley and Herself*. Watertown, MA: Eastgate Systems, 1995.

Jameson, F. *Postmodernism or, The Cultural Logic of Late Capitalism*. London: Verso, 1991.

Jurgenson, N. 'Digital Dualism Versus Augmented Reality'. *Cyborgology*. 24 February 2011. Web. 28 January 2019. https://thesocietypages.org/cyborgology/2011/02/24/digital-dualism-versus-augmented-reality/.

Kearney, R. *On Stories*. London: Routledge, 2002.

Kearney, R. *Poetics of Imagining: Modern to Post-Modern*. Edinburgh: Edinburgh University Press, 1998.

Kearney, R. *The Wake of Imagination*. London: Taylor and Francis, 1988.

Kirby, A. *Digimodernism: How New Technologies Dismantle the Postmodern and Reconfigure Our Culture*. London: Continuum, 2009.

Koehler, A. *Composition, Creative Writing Studies, and the Digital Humanities*. London: Bloomsbury, 2017.

Konstantinou, L. 'Four Faces of Postirony'. *Metamodernism: Historicity, Affect, and Depth After Postmodernism*. Eds, van den Akker, R. Gibbons, A. and Vermeulen, T. London: Rowman & Littlefield International, 2017, 87–102.

Lambert, J. *Digital Storytelling: Capturing Lives, Creating Community*. London: Routledge, 2013.

Landow, G. P. *Hypertext: The Convergence of Contemporary Critical Theory and Technology*. Baltimore: John Hopkins University, 1992.

Latour, B. *Down to Earth: Politics in the New Climatic Regime*. Cambridge: Polity Press, 2018.

Lerner, B. *10:04*. London: Granta, 2015.

McClanahan, A. *Dead Pledges: Debt, Crisis, and Twenty-First Century Culture*. Stanford: Stanford University Press, 2017.

Miéville, C. *The City & the City*. London: Macmillan, 2009.

Miéville, C. 'Reports of Certain Events in London'. *Looking for Jake and Other Stories*. London: Macmillan, 2005, 53–75.

Montuori, A. 'Foreword: Transdisciplinarity'. *Transdisciplinarity: Theory and Practice*. Ed, Nicolescu, B. Cresskill, New Jersey: Hampton Press, 2008, ix–xvii.

Moraru, C. *Cosmodernism: American Narrative, Late Globalization and the New Cultural Imaginary*. Ann Arbor, Michigan: The University of Michigan Press, 2011.

Moulthrop, S. 'Rhizome and Resistance: Hypertext and the Dreams of a New Culture'. *Hyper/Text/Theory*. Ed, Landow, G. P. Baltimore: John Hopkins University Press, 1994. 299–319.

Nicolescu, B. '*In Vitro* and *In Vivo* Knowledge – Methodology of Transdisciplinarity'. *Transdisciplinarity: Theory and Practice*. Ed, Nicolescu, B. Cresskill, New Jersey: Hampton Press Inc.: 2008, 1–21.

Olsen, L. and Guthrie, T. *10:01*. Electronic Literature Collection, Vol. 1, 2005. Web. 20 March 2018. http://collection.eliterature.org/1/works/olsen_guthrie__10_01/1001.html.

Payam, M. S. Internet and Digital Textuality: A Deep Reading of *10:01*'. *The Bloomsbury Handbook of Electronic Literature*. Ed, Tabbi, J. London: Bloomsbury, 2018, 311–321.

Peck, J. 'The Cult of Urban Creativity'. *Leviathan Undone? Towards a Political Economy of Scale*. Eds, Keil, R. and Mahon, R. Vancouver: University of British Columbia Press, 2009, 159–176.

Pope, R. *Creativity: Theory, History, Practice*. London: Routledge, 2005.

Pullinger, K. *Breathe*. Editions at Play, 2018. Web. 5 February 2018. https://editionsatplay.withgoogle.com/#!/detail/free-breathe.

Ricardo, F. J. 'Juncture and Form in New Media Criticism'. *Literary Art in Digital Performance: Case Studies in New Media Art and Criticism*. Ed, Ricardo, F. J. London: Continuum, 2009, 1–9.

Ricoeur, P. *Time and Narrative, Volume 3*, Trans. Blamey, K. and Pellauer, D. Chicago: University of Chicago Press, 1988.

Ryan, M., Foote, K. and Azaryahu, M. *Narrating Space / Spatializing Narrative: Where Narrative Theory and Geography Meet*. Columbus: Ohio State University Press, 2016.

Salami, M. *John Fowles' Fiction and the Poetics of Postmodernism*. Vancouver: Fairleigh Dickinson University Press, 1992.

Smith, H. and Dean, R. T. 'Introduction: Practice-led Research, Practice-led Practice – Towards the Iterative Cyclic Web'. *Practice-led Research, Research-led Practice in the Creative Arts (Research Methods for the Arts and Humanities)*. Eds, Smith, H. and Dean, R. T. Edinburgh: Edinburgh University Press, 2009, 1–38.

Smyth, K., Power, A. and Martin, R. 'Culturally Mapping Legacies of Collaborative Heritage Projects'. *Valuing Interdisciplinary Collaborative Research: Beyond Impact*. Eds, Keri Facer and Kate Pahl. Bristol: Polity Press, 2017. 191–213.

Spiers, H., Hornberger, M., Bohbot, V., Dalton, R., Hölscher, C., Manley, E., Sami, S., Silva, R. and Weiner, J. Sea Hero Quest. London: Glitchers, 2016.

Steiner, G. *Grammars of Creation*. London: Faber and Faber, 2001.

Tabbi, J. 'Introduction'. *The Bloomsbury Handbook of Electronic Literature*. Ed, Tabbi, J. London: Bloomsbury, 2018, 1–9.

Thumim, N. 'Therapy, Democracy and the Creative Practice of Digital Storytelling'. *Digital Storytelling: Form and Content*. Eds, Dunford, M. and Jenkins, T. London: Palgrave Macmillan, 2017. 229–240.

Tringham, R. 'Creating Narratives of the Past as Recombinant Histories'. *Subjects and Narratives in Archaeology*. Eds, van Dyke, R. M. and Bernbeck, R. Boulder, Colorado: University Press of Colorado, 2015. 27–54.

Tuer, D. *Mining the Media Archive: Essays on Art, Technology And Cultural Resistance*. Toronto: Y Y Z Books, 2006.

Turchi, P. *Maps of the Imagination: The Writer as Cartographer*. San Antonio, TX: Trinity University Press, 2004.

Walsh, J. *Seed*. London: Editions At Play, 2017. Web. 5 February 2018. https://seed-story.com.

Webb, J. *Researching Creative Writing*. Newmarket, Suffolk: Frontinus, 2015.

4 Hypertextual adventures

Introduction

Michael Joyce's seminal first-generation hypertext, *afternoon: a story* (1987), published on the Storyspace platform, tells the story of Peter, a technical writer, who has recently divorced his wife, Lisa. After witnessing a car accident, Peter becomes convinced that both his wife and son, Andy, have been killed in the crash. He starts to blame himself for not going to help with the crash. Peter becomes ever more neurotic in his thoughts. Dream and reality converge as other characters, including Wert, a work-colleague and Peter's ex-wife's lover, and Lolly, Peter's secretary and lover, drift in and out of the story. When he returns to the accident there are only tyre marks and some old papers.

> Most of the papers are old and waterstained, dried by the sun into yellowing things. There is a fresh white paper with my son's name upon it, and red markings from a teacher. It is a report on Louis Quattorze, and his looping handwriting makes me weep. It begins: "I am the Sun King," said Louis the Fourteenth of France.
>
> (Fenceline)

The child's homework might be Andy's, indicating he was involved in the accident. Or it could simply have been dropped by another child. We are never told for sure either way. This lack of closure, or narrative certainty, is one of the key features of the story. The other is its fundamental multi-linearity. The story has 539 lexias and 951 separate links and is designed so that no individual pass through will be the same. The availability of some lexias are conditional on the reader having reached a specific subset of other lexias (using the 'guard' function in Storyspace). J. Yellowlees Douglas estimates that most readers will visit less than ten per cent of the lexias in the text after their first reading. The navigational architecture is minimal. There are 'Yes' and 'No' buttons at the bottom of the window. However, hyperlinks within each lexia are hidden; the user is forced to search for them by clicking on words and images: 'the story exists on various levels and changes according to chosen paths, such that even previously visited pages change in terms of designated and hidden links' (Ensslin 2007, 70). These effects are amplified by the first-person unreliable voice through which the stream of consciousness is delivered. If there is any 'ending' to the story, then perhaps it comes (or a version of it comes) when the reader finally reaches the 'white afternoon' lexia, something that is not straightforward, given the complex series of 'guard fields' that surround it. Once reached, however, the reader discovers that Peter may have been the driver of the car in which his wife and son were killed. The ambiguity within the story is suddenly revealed to be (possibly) a form of Peter's denial of his guilt. As Hayles notes in her own close reading of the story, '[Peter] does not want to face what in some sense he already knows' (2008, 61). Yet this is only one possible ending; there is no guarantee that the reader will get to this lexia in any single pass through. Since there is no overview map or structure given to the reader, they do not know what they have not seen. Yet, even if 'white afternoon' is reached, the overall effect of the story is to destabilise the reading experience in multiple ways, denying the reader coherency of plot and meaning.

For these reasons, the work has been described as postmodernist in its design; excerpts from it were included in the 1998 edition of the *Norton Anthology of Postmodern Literature*. Ensslin has gone as far as to say that 'Joyce's central intention is to illustrate how hypertext embodies postmodernist writing and poststructuralist theory' (2007, 70).

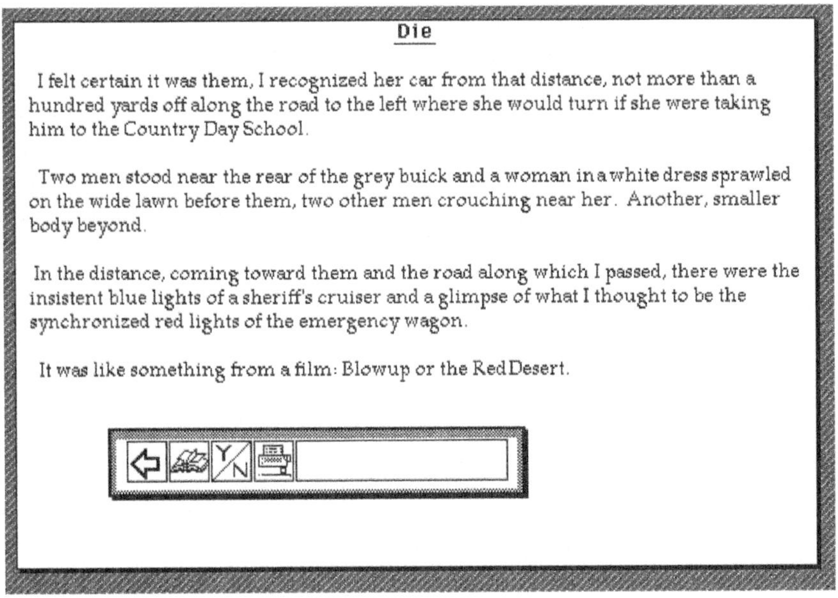

Figure 4.1 Screenshot showing Die Lexia of *afternoon: a story* (1987).
Source: © Michael Joyce.

It is clear that, in writing *afternoon: a story*, Joyce was self-consciously experimenting with hypertextual form and agency. In his book, *Of Two Minds: Hypertext, Pedagogy, and Poetics* published in 1995, Joyce outlines what could be termed his approach to hypertext fiction at this time. Written at the high point of first-generation hypertext, it encapsulates what might now be viewed as a utopian belief in hypertext's destiny to replace print-based media. In a chapter entitled 'What I really Wanted to Do I Thought', Joyce passionately writes about how the computer finally allowed him to satisfy his artistic dream of being able to write a novel that changed every time someone read it (1995, 32). Elsewhere in the book, he goes on to differentiate between what he calls *exploratory* and *constructive* hypertexts. Exploratory hypertexts are navigational in purpose, helping the reader find and collate information. Although the reader constructs their own reading by selecting what sources to access, they remain at some distance from the text; the role of reader and author remain distinct. With constructive hypertext, however, the reader can actively link between and among texts rather than following predetermined pathways, thus creating new pathways. As Joyce writes:

> Constructive hypertexts ... require a capability to act: to create, to change, and to recover particular encounters within the developing body of knowledge... These encounters, like those in exploratory hypertexts, are maintained as versions, i.e., trails, paths, webs, notebooks, etc.; but they are versions of what they are becoming, a structure for what does not yet exist.
>
> ('Siren Shapes' 1993, 42).

Most hypertext fictions come under Joyce's exploratory category, including *afternoon: a story* as well as a lot of other notable work such as Shelley Jackson's *Patchwork Girl* (1995). Exploratory hypertexts operate in the same way as hard-copy books, in the sense that they are stable, with a clear separation between readerly and authorial roles. As Scott Rettberg states, 'the reader may explore, mark, and make annotations to an exploratory hypertext, but in doing so the reader is not modifying the work itself' (2011, 189). In *afternoon: a story* there are no links to any external site or content; the author's control is overt and explicit in that the reader is effectively forced to follow particular pathways, creating, what Hayles terms, 'an oppressive sense of being required to jump through the same series of hoops numerous times' (2008, 62). It is for this reason that Hayles calls *afternoon: a story* printcentric in its design, in other words a story firmly situated in the print-based traditions of narrative.

Of course, by the time Hayles was writing about the story, those early years of hypertext fiction were disappearing into the middle distance. Captured in her comments are not just an acknowledgement of the limitations of the story but also the belief of those first-generation hypertext authors in the revolutionary impact of what they were doing. Although Joyce has backtracked to some degree, in the late 1980s and early 90s there was still very much a belief that hypertext fiction was offering something radical and new, transforming the relationship between reader and writer. In 'Hypertext and Hypermedia', an essay first published in 1993, Joyce eulogised that hypertext readers alter the form of what they read by the choices they make and thus 'in a very real sense write (or rewrite) hypertexts' (1993, 20). In hypertext fiction, the reader thus became a new kind of entity, the like of which had not been seen before. George Landow christened this new entity a 'wreader' (1992).

For Joyce, a constructive hypertext, on the other hand, is a far-more complex formulation. Here, a predefined narrative structure is minimised. Instead, the emphasis is on the reader's role in the construction of any story or narrative pathway. Rettberg notes that these constructive hypertexts can either be individually written, or written collectively, 'in which case a community of reader/writers are actively interacting with, forging connections, expanding upon, and reacting to the work of others' (2011, 190). A constructive hypertext, then, can be understood as a participatory writing performance, what Rettberg calls an 'event' as much as a 'work' (190). A good of example of this kind of project is the collaborative hypertext novel *The Unknown* (1998). Although the primary authors are identified as William Gillespie, Scott Rettberg and Dirk Stratton, a variety of other authors and artists participated over a period of four years between 1998 and 2002 in what finally became the novel. The basic premise of *The Unknown* was to write a satirical hypertext about a book tour.[1] Other than that, it was pretty much down to the individual contributors in terms of what was actually written and how those materials were linked within the developing architecture of the work. The finalised novel consists of a great variety of texts, including what appears to be authentic cover letters and transcripts of conversations (in other words paratexts about the novel), alongside elements from the novel itself. The authors call the work 'encyclopaedic'; indeed, *The Unknown* exists as a vast sprawl of hypertext pages through which the reader navigates their way. As such, there is no overall narrative arc or character development; on top of this, it was recognised that the reader would only ever access a tiny part of the material. Consequently, the authors recognised the need for each page or narrative element to function individually, at least to some degree (Rettberg 2011, 193). Some of the writing was done collaboratively, in real time, where different authors would take it in turns at the keyboard; other scenes were written by a single author working in isolation. Yet, crucially, all the participants in the novel knew each other, and therefore formed, what can be understood as, a closed community. For Scott, it was the fact that the authors had prior knowledge of each other as people and artists that made the writing process such a success. As he notes, '[w]ithout these preexisting relationships and ongoing negotiations about the shape of the story, the project would not have come to pass' (2011, 193).

Underlying the entire project, however, is something far-more fundamental: postmodernism. Like many of the hypertextual projects of this

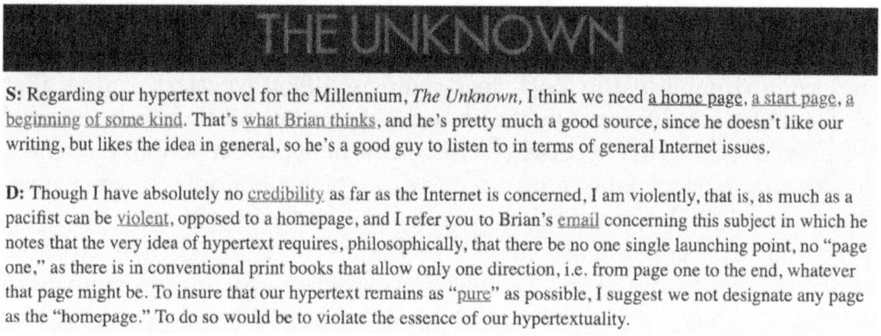

THE UNKNOWN

S: Regarding our hypertext novel for the Millennium, *The Unknown,* I think we need a home page, a start page, a beginning of some kind. That's what Brian thinks, and he's pretty much a good source, since he doesn't like our writing, but likes the idea in general, so he's a good guy to listen to in terms of general Internet issues.

D: Though I have absolutely no credibility as far as the Internet is concerned, I am violently, that is, as much as a pacifist can be violent, opposed to a homepage, and I refer you to Brian's email concerning this subject in which he notes that the very idea of hypertext requires, philosophically, that there be no one single launching point, no "page one," as there is in conventional print books that allow only one direction, i.e. from page one to the end, whatever that page might be. To insure that our hypertext remains as "pure" as possible, I suggest we not designate any page as the "homepage." To do so would be to violate the essence of our hypertextuality.

Figure 4.2 Screenshot showing Index Lexia of *The Unknown* (1998).
Source: © William Gillespie, Scott Rettberg and Dirk Stratton.

period, *The Unknown* is in essence an exploration of postmodern creativity. The deep irony, fragmentation and depthlessness of the work, the satirical use of pastiche, are all classic tropes of the postmodern genre, of course. Even the title of the work itself – *The Unknown* – speaks of the ultimate unknowability that haunts the postmodern condition. As we're told on the very first page (Figure 4.2) after a quote from Thomas Pynchon, the high-priest of postmodernity: 'What you don't know can and will hurt you, but not as much as will what you already know which already has and will continue to bring you pain' {index}. *The Unknown* is therefore an attempt to map the political, social and cultural contours of our ignorance, the structures of power that lie behind knowledge as a fictive entity. The labyrinth structure of the work reflects the self-referential nature of language; indeed, the shadow of Jorge Luis Borges falls heavily across the entire project. As the authors openly admit in the story itself, '*The Unknown* could not exist without [Borges]. We're all still following the paths laid through his many Labyrinths' {borges}.

The publication of *afternoon: a story* (1987) and *The Unknown* (1998) spans twelve years, a small period of time yet a critical one nevertheless, which saw the rise of hypertext fiction to become *the* creative form of digital storytelling (Landow 1992). As I've shown, while *afternoon: a story* follows the format of what Joyce termed exploratory hypertexts, *The Unknown* is more ambitious in design and narrative purpose, embracing Joyce's second category, constructive hypertexts. Yet *The Unknown* comes right at the end of this period of hypertextual dominance and is curious in the way it both looks forward to the participatory storifying of the social media age (discussed in Chapter 7), and backwards to the avant-garde experimentalism of the classic hypertext novel. For the authors of both these works, however, the hypertext story remained a powerful medium by which a postmodern ontology could be actuated, foregrounding the constructed and partial nature of knowledge.

Yet, as we've seen in Chapter 2, the modality of the labyrinth (to return to Kearney's schematic), was not to last. In fact, postmodernity was well and truly on its last legs by the time *The Unknown* was published in 1998. As I've explained, the new period that subsequently emerged has no agreed name, although some academics have labelled it metamodern (van den Akker, Gibbons and Vermeulen 2017). As a creative modality, I have described it as a period of the map/rhizome/string-figure. I have also made the point that this new creative modality has re-envisioned the concept of postdigitality; indeed, the previous chapter argues that the postdigital is a fundamental part of our map/rhizome/string-figure modality.

This enmeshment between technology and 'ways of seeing and representing the world' is a fundamental one, of course. Although it is not the purpose of this chapter to explore the history of this relationship in any great detail, it is, nonetheless, important that the postdigital is understood as but the latest phase within a much longer continuum. From this perspective, hypertext stories such as *afternoon: a story* and *The Unknown* offer insight into not only the technology of the time, but also the way technology operates symbiotically with social, cultural and economic norms. As David Trotter says, '[c]ommunications technology is an attitude before it is a machine or a set of codes' (2013, 2). And it is such attitudes, customs and habits that creative practitioners draw on in their representations of the world. Just to be clear, then, this chapter is not simply focussed on the direct, physical, relationship between technology and creative practice, in other words the creative affordance that certain technologies (such as hypertext) gives to artists and authors. Rather, it is also interested in broader and deeper phenomena of which technology is but one intimately connected part. These are phenomena related to epistemology and ontology and relate to fundamental questions concerning existence and reality. I would argue that it is only by a consideration of these wider issues that a proper understanding of postdigital poetics, praxis and research can be achieved. As Sara Danius puts it, '[t]echnology helps change not only the world but also the perception of that world' (2002, 189).

In its analysis of these issues, this chapter attempts to do two things. First, it seeks to provide an overview of the relationship between technology and storytelling from the late nineteenth century onwards, in particular noting the impact that various technologies had on the stylistic approaches of various authors. As Trotter notes, some of the issues raised by technology today were first highlighted in the years immediately before the Second World War. Second, it places this relationship within the paradigmatic shifts in ontology that I've already introduced, namely the mirror, lamp, labyrinth, map/rhizome/string-figure schematic. Central to this is an in-depth analysis of the development of hypertext fiction as the epitome of postmodernity. The following chapter then details how these hypertextual labyrinths provided the foundations on which postdigital storytelling rests by looking at where hypertext fiction is today. As Joyce himself ends the introduction to *Of Two Minds*, in the end all that matters is the human need for connection, 'to penetrate the dark edge of existence comforted by knowing that we are not lost to one another' (1995, 15).

As I've explained, the focus of this book is the impact of postdigitality on digital storytelling: storytelling that is purely within the digital domain but also stories that are between the domains too (the degree to which digital storytelling embraces non-digital elements). This is the *within* or *across* the digital/non-digital nexus that I spoke about in Chapter 3. Although the focus of this, and the following chapter remains digitality (in this case hypertext storytelling), the wider implications of our 'within' and 'across' parameters continue to inform the discussion. What I mean by this is that, in our exploration of the digital/non-digital nexus, I'll also be analysing the impact that hypertextuality had on printed text and transmedial hybridity more generally.

In terms of my definition of hypertext storytelling, I draw on Astrid Ensslin's definition of hypertext in her book, *Canonizing Hypertext: Explorations and Constructions* (2007, 5–7). Like Ensslin, my focus is on textual works (prose), rather than work that is primarily (if not completely) visual or aural (video, film, animation etc.), although these textual works will very often be multimodal in nature, of course. I also do not include works that are primarily game-orientated in design and purpose, although I recognise that ludic design continues to have a significant influence on literary work more broadly (see Ensslin, *Literary Gaming* 2014).

Storytelling and technology

As I've already stated, it's not the purpose of this book to provide a detailed historical account of the relationship between storytelling and technology. The focus of my analysis remains very much on where we are today in terms of storytelling. However, a book that has the prefix 'post' in its title cannot eschew historicity in its entirety. By definition, 'post' not only suggests a 'before' but also implies some kind of explanatory relationship between the two positions, whereby 'what came after' can only be understood through a consideration of 'what went before'.

The primary focus of this and the next chapter is hypertext fiction, the sort of digital storytelling that Ruth Page and Bronwen Thomas describe as reaching 'canonical status' by the end of the 1990s (2011, 1). If we're to properly understand postdigital storytelling and its relationship to the map/rhizome/string-figure modality, then we also need to understand how stories such as *afternoon: a story* and *The Unknown* were expressions of the prior modality of postmodernity. Understanding this relationship, the characteristics of and the reasons for, the paradigmatic shift from labyrinth to map/rhizome/string-figure are critical to what I'm going to be doing here.

As William Uricchio notes, media should not be understood as simply technologies, institutions or texts, but rather as a much wider set of 'cultural practices' within which are entangled the 'lived experiences' of those who produce, define and use them (2003, 24). It is how such 'lived experiences' are reflected in the creative output of artists and writers that is at issue here. In other words, we need to comprehend how, what might be termed, the aesthetics of representation were transformed by technological change. We then also need to consider what this means for storytelling generally as a form of research.

Richard Menke, in his book, *Telegraphic Realism: Victorian Fiction and Other Information Systems* (2008), examines how 'imaginative writing' responded to the 'Victorian information technologies' of the Penny Post, electric telegraph and wireless telegraphy (3). Critical to his study is the way the modality of 'realism' grappled with these transformations. In his analysis of the electric telegraph, for example, Menke notes how both Elizabeth Gaskell and Charles Dickens used the metaphor of the telegraph to describe the way their fiction worked: both possessed the ability of being able to neutrally transmit knowledge of a shared reality.

> When Gaskell and Dickens consider the electric telegraph as a model for fiction, they treat it in similar ways: as a vehicle of narrative truth-telling, a figure of fictional connectivity, and a confirmation of the meaningful ties that wire together the world.
>
> (2008, 88)

For Menke, then, the new media technologies of the nineteenth century were understood and represented through the modality of realism (Kearney's mirror). Here, like the very act of writing itself, information systems were seen as neutral transmitters of truth. Yet, as a set of 'cultural practices', such technologies also heralded a far-more elusive, though none-the-less important, transformation. How individuals and communities responded to this 'new media ecology', with its faster and broader information flows and social interconnectedness, is a key aspect of artistic representation in this period. Authors such as Gaskell, Dickens and Eliot, through works that included *Cranford* (1853), *Hard Times* (1854) and *Middlemarch* (1871), became fascinated by the representation of inner, private thoughts that the Penny Post and telegraphy now made available. Equally, this new interconnectedness, running through and across entrenched class divisions, allowed writers to reconceive social and cultural dynamics in ways that were both new and innovative (Goble 2010, Chapter 1).

Yet there was also a tension in the work of these writers, an implicit understanding that these new forms of media offered a unique threat to the status and rationale of the literary novel. As Menke notes, '...we should think of Victorian realism itself as an exploration of the power and the limits of written textuality in an age busy producing alternatives to it' (2008, 11). And this sense of threat, of the need for artists to engage directly with the radical implications of technological change, continued on into the twentieth century. Danius calls this a 'crisis of the senses' (2002, 3), a term she uses to describe the ways in which perception was fundamentally reconfigured by late-nineteenth and early twentieth-century technological innovations, including phonography, cinematography, telephony, electricity and the motor car. This 'crisis' became a key factor in the development of modernist aesthetics; in other words, technology was not just the context for modernism but rather a critical constitutive element (see also Goble 2010).

Modernism is a notoriously imprecise term and it certainly isn't the purpose of this chapter to delve too deeply into the debates around its definition. However, I would argue that understanding some of the main arguments concerning modernist aesthetics and technology is important if we are to more fully comprehend what actually is going on in our own period. One of modernism's key characteristics, for example, was its rejection of nineteenth-century realism. In other words, by the early twentieth century, technology had played a significant part in pushing artists away from a simplistic understanding of how the world is experienced and understood. Central to this were the innovative ways in which 'sense data' could now be technologically stored, transmitted and reproduced. The problem for artists, then, was 'how to represent authentic experience in an age in which the category of experience itself [had] become a problem' (Danius 2002, 3). This 'authentic experience' was no longer to be found with the nineteenth-century omniscient author or with novels whose range and depth attempted some kind of objective totality. Instead, 'the crisis of the senses' foregrounded human subjectivity as the fundamental epistemological position, in which consciousness, perception, emotion and meaning were prioritised. In terms of literature, writers such as Virginia Woolf and James Joyce experimented with new forms and techniques, including interior monologue, stream of consciousness and defamiliarisation. If there was a realism still to be grasped through artistic expression, then it could only come from a recognition that all perception was ultimately a subjective experience. In this sense, the reader's task was to complete the work of art, to decode the unfamiliar, and, in that way, foreground the authentic process of conscious understanding. Woolf's short story, 'The Mark on the Wall' (1919), is a good example of this, consisting entirely of a stream of consciousness that captures fleeting thoughts and sensory impressions:

> O dear me, the mystery of life! The inaccuracy of thought! The ignorance of humanity! To show very little control of our possessions we have–what an accidental affair this living is after all our civilization.
>
> (2008, 4)

Both the theme and form of Woolf's short story reflect the avant-garde experimentalism at the heart of modernist aesthetics. As Adrian Hunter says, Woolf's concerns were twofold: first, she was interested in the way the mind experienced reality; and second, she strove to understand how best the writer could convey that experience in narrative form (2007, 63). Yet it was also this avant-gardism, the belief in a cultural elite, that became another important characteristic of the modernist movement.

Not all writers in this period were quite so emollient in their response to technology. The development of the cinema, radio and phonograph were seen by some as a serious threat to the predominance of literature, and through that, to enlightened culture itself. In an article written for *Scribner's Magazine* in 1894, Octave Uzanne, the French writer, publisher and journalist, relates a conversation over dinner at the Royal Society in London at which the guests take turns in offering predictions on the state of the world by the end of the twentieth century. When it comes to Uzanne's turn, he speculates that the book will be 'threatened with death by the various devices for registering sound which have lately been invented' (1894, 224). In both a lighthearted and satirical article, Uzanne predicts a world dominated by the phonographic novel, in which illustrations and live performance are transferred to 'the Kinetograph of Thomas Edison' in which 'scenes described in works of fiction and romances of adventure will be imitated by appropriately dressed figurants and immediately recorded' (1894, 229). As one of the guests notes at the very end of the piece, heavily laced with sarcasm, '...but what happiness not to be obliged to read [books], and to be able at last to close our eyes upon the annihilation of printed things!' (1894, 231).

For Walter Benjamin, however, it was precisely the disruptive potential of technology that was to be celebrated. In his essay 'The Work of Art in the Age of its Technological Reproducibility' (1936), Benjamin argues that the most significant change introduced by new technology is what he terms the 'massification' of art, in other words, the increased accessibility to artistic output by those previously shut out or ignored by the avant-garde. As a committed socialist thinker and philosopher, Benjamin believed that the breaking down of elitism and privilege wherever it existed was a human obligation. From this perspective, the new media technology of the late nineteenth and early twentieth century was perceived as inherently egalitarian and liberating. Benjamin argued that a technology that allows for the mass reproduction of an oil painting undermines the individual 'aura' of the original work of art and instead 'substitutes a mass existence for a unique existence' (2002, 104). In his essay 'The Author as Producer' (1934), Benjamin makes the case for literature to learn from modernist new media and to transform the passive reader into an active participant, and in that way release the working classes from economic and social subjugation. He goes on to suggest a rubric by which to appraise all new literary forms: 'this apparatus is better the more consumers it is able to turn into producers – that is, readers or spectators into collaborators' (1996, 777). The inherent benefit of new technology was its ability to disrupt the ivory towers of the avant-garde and by doing that universalise both who can write and what they can write about: 'Authority to write is no longer founded in a specialist training but in a polytechnical one, and so becomes common property' (1996, 778).

Virginia Woolf was the sort of avant-garde writer that Walter Benjamin wished to see overthrown by new technology. Yet, as I've already shown, for modernist writers such as Woolf, technology elicited a huge wave of innovation in their own work. In her essay 'Poetry, Fiction, and the Future' (1927), Woolf argues that new media have created a paradox of connectedness, linking citizens over huge distances through the technology of 'telepresence' yet at the same time alienating them from their closest neighbours. Woolf's response to this new condition was to innovate. If the telephone and radio helped foster a 'mind…full of monstrous, hybrid, unimaginable emotions' (1927, 429), then poets and authors must develop hybrid genres to represent this new form of modernity.

Trotter calls this interwar period in Britain 'the first media age', witnessing, as it did, technological innovations such as the telephone, phonograph, film and the cheap rotary press, as well as new hybrid media such as broadcast radio and television (2013, 8). Alongside this, wireless telegraphy together with advances in air, road and rail technology, became key building blocks of a hitherto unimaginable global connectivity. All of these unprecedented developments led to what Trotter calls a 'meta-attitude' that was a direct response to 'the proliferation of connective uses of media' (2013, 273). And it was this meta-attitude that played out across avant-garde literature of the time.

Telephony, for instance, clearly made a significant impression on modernist writers such as Woolf. Behind her concept of 'telepresence' lay, what Trotter terms, a new kind of 'connective sociability' (2013, 38). Figure 4.3 shows the way this new connectivity was advertised to the 'Telephone-Wife' of the day. By 1937 the General Post Office in the UK had celebrated the installation of a million phones in London alone, with three million nationwide (Trotter 2013, 56). For the first time in human history instantaneous conversation could be conducted at distance. For the middle classes, these intimate, private liaisons occured in the privacy of the home and could transgress entrenched social and sexual barriers. In other words, the technology facilitated (at distance) sociability while at the same time enforcing new forms of (domestic) privacy. Literature seized on the narrative possibilities of this paradox. Key here was a more sophisticated and overt use of dialogue as a narrative device. It was in this period, for example, that conversation, driven by the proliferation of telephony, started to replace more traditional forms of communication, such as letter writing and telegram. As David Lodge states, this had a singular effect on the way dialogue was used in the modernist novel:

> The stream of consciousness gives way to a stream of talk, but it is talk without the reassuring gloss of the classic novel's authorial voice, without a privileged access to the thoughts and motivations of characters, so that the "modern" note of disillusion, fragmentation and solipsism persists.
>
> (1990, 81)

WONDERS OF THE TELEPHONE

III.—The Happy Eyes of the Telephone-Wife

They twinkle and glow, those clear bright eyes of hers, because they are the windows through which all the sunniness of her nature shines out —unclouded by the worries of domestic life.

For her, one simple contrivance has reduced those worries to insignificance.

She is the Telephone-Wife, the wife for whom a thoughtful husband (or her own mother-wit) has provided a telephone. It is her ally and helpmeet to cope with harassing occasions.

That womanly feeling of loneliness! Thanks to the telephone, husband or friend is ready at hand.

That fear of burglars at night, or of fire, or of sudden sickness! Down the telephone lies the quickest path to aid.

For those workaday affairs— whether with butcher or baker or candlestick-maker; for social occasions, foreseen or impromptu, whose going-right means pleasure and wifely pride, whose going-wrong means worry and the shame of failure —the telephone is her standby.

Her servant in big things and little, a breaker of isolation, a leaper of distances, a link with powers greater than her own — the telephone gives her that sense of care-freedom which her cheery eyes reflect.

TDA

Issued by THE TELEPHONE DEVELOPMENT ASSOCIATION, Aldine House, 10-13, Bedford Street, London, W.C. 2.

Figure 4.3 'Wonders of the Telephone' (1925). Telephone Development Association flyer.
Source: © Illustrated London News Ltd/Mary Evans.

A good example of this is Evelyn Waugh's *Vile Bodies* (1930) in which an entire chapter is given over to the telephone conversation between Adam Symes and his girlfriend Nina. The dialogue is written almost entirely without authorial intervention, the dramatic irony heightened by the partiality of each character's point of view. Repetition is a key technique here, the repeated use of 'I see' merely re-emphasising the misunderstandings between the characters. As Bronwen Thomas notes, Waugh skillfully suggests both the emotional and physical distance between the characters; he also hints at the inner fragility of these bright young things, where the telephone has become a means of guarding against unwanted emotional responsibility (2012, 134).

This 'talk explosion' (Kacandes 2001) had more subtle effects too. Trotter, for instance, argues that the rise of the 'collective novel' in the 1930s was, in part, a response to the growth of cinema as a new media form, particularly the talkie. Films such as *Grand Hotel* (1932) created 'an opportunity to imagine a particular kind of talk as a moral and political value, and to incorporate it into a social logic of space: an understanding of coexistence, of cross-class cooperation, of community, even' (2013, 200). Such 'collective novels' were firmly situated on the political Left; they were 'collective' not because they were written by more than one person, but rather that they tried to represent society as an integrated, collective whole: the focus of the collective novel was less the life of a single character than the social and moral interconnectedness of men and women. Harold Heslop's *Last Cage Down* (1935) represents the life of a mining community in the North East of England, in which the struggle between workers and management remains the primary focus. The idiom of the talkies, with its straight-talking American slang and informality, is used by the working-class characters both as a political tool and as a form of collective identity, undercutting the rigid formality of the (middle-class) managers: 'Go on, Frost, shoot your mouth off, as the pictures say. Pour it all out. It doesn't worry me' (1935, 85).

After the Second World War, the pace of technological innovation continued to accelerate. Exactly when modernism slipped into postmodernism is a moot point; for Frederic Jameson, the critical conditions of what he termed 'a new economic world system' were already in place by the 1950s (1991, xx). What this amounted to was a new kind of global connectivity in which the individual was reduced to a single node within a vast network of capital. Alongside these economic developments came technological innovations such as television and networked computer systems, together with the increasing global spread of radio, cinema and cheap print-based media. According to Jean Baudrillard, these media-based technologies ensured that existence was now pure simulacrum, caught up in globalised networks of data and capital, with no meaningful relationship to reality (1983). And all this was happening within the context of decolonisation with its erosion of the social, cultural and economic certainties of empire.

As David Harvey notes, the ideals of the Enlightenment were still very much at the core of modernism: a belief in positivism, in linear progress and absolute truths, and ultimately, in the ability of an educated human mind to perceive and understand reality (1990, 9). These ideals were certainly under attack by the 1930s, as I've already shown; yet as the postwar period lengthened, these certainties suffered existential attack. In their place came a far-more radical set of values, privileging fragmentation, indeterminacy and an 'intense distrust of all universal or "totalizing" discourses' (Harvey 1990, 9). For Brian McHale, modernist fiction was essentially epistemological in its focus, asking questions about the nature and limits of knowledge (1999, 9). Postmodern fiction, on the other hand, is described as being ontological, in other words, work that foregrounds questions around the nature of existence and hermeneutics (10). As McHale goes on to say: '... postmodernist fiction ... is above all illusion-breaking art; it systematically disturbs the air of reality by foregrounding the ontological structure of texts and of fictional worlds' (221). One of the great examples of this is Thomas Pynchon's *Gravity's Rainbow* (1973). According to McHale, in his book, *Constructing Postmodernism*:

> Pynchon's readers have every right to feel conned, bullied, betrayed. Indeed, these responses are the essence of the aesthetic effect of *Gravity's Rainbow*. We have been invited to undertake the kinds of pattern-making and pattern-interpreting operations which, in the modernist texts with which we have all become familiar, would produce intelligible meaning; here, they produce at best a parody of intelligibility. We have been confronted with representations of mental processes of the kind which, in modernist texts, we could have relied upon in reconstructing external (fictive) reality. In *Gravity's Rainbow*, such representations are always liable to be qualified retroactively as dream, fantasy, or hallucination, while the reconstructions based upon them are always subject to contradiction or cancellation. The ultimate effect is radically to destabilize novelistic ontology.
>
> (1992, 81)

The structure of *Gravity's Rainbow* is labyrinthine, involving over 400 characters in a disconcerting mashup of literary fiction, pornography, slapstick and technical scientific material. Like a labyrinth, it lures readers 'into interpretive dark alleys ... requiring them to find their way out by some other path than the one they came in by' (1973, 82). Yet the novel can also be seen as a response to, what Luc Herman and Steven Weisenburger call, 'the ceaseless, ever-morphing needs of a corporatized military-industrial management and production regime' (2013, 26). In other words, at the heart of *Gravity's Rainbow* is a concern with Jameson's notion of late capitalism and, in particular, the way everyday existence was increasingly mediated through globalised technological networks. And of course, it was such technological developments that in turn led to what is considered to be the first significant phase in digital storytelling: hypertext fiction.

Hypertext fiction: beginnings

I've attempted to do a number of things in the preceding section. I've outlined some of the key characteristics of both modernism and postmodernism, two philosophical and cultural movements that together dominated the twentieth century. In particular I've shown how postmodernism was, in many ways, an overt rejection of modernism's key tenets. Yet, what I've also shown, is that both movements were intricately bound up with the technological and economic developments over this period. By the early twentieth century, innovations as diverse as telephony, cinema, radio and the motor car, had begun to radically transform how individuals engaged with each other and the wider world. And by the interwar period, Western European and American societies were experiencing a new kind of technologically-mediated connectivity, the effects of which amounted to the dawning of a new media (or mediatised) age.

All this change had a powerful impact on how artists and writers began to think about the nature and role of their work in the representation of, what Danius labels, 'authentic experience' (2002, 3). Trotter calls this a literary meta-attitude and it had at least two elements. The first was a response to the social and cultural changes taking place: for example, in the way the telephone or the cinema began to change every day behaviour. The second was perhaps less obvious but just as important: the quest to find a role for literature in an increasingly mediatised society. Quite simply, modernists such as James Joyce and Virginia Woolf saw their task as much about seeking to find the unique affordance of written textuality in the face of new media proliferation, as it was about reflecting these changes within their own writing. Crucially, then, technology was not just a social condition; it also enacted new and transformative ways of thinking about existence and the nature of reality (the epistemology described by McHale).

As I've shown, these factors continued to proliferate at an increasing pace after the Second World War. Postcolonial independence and the decentering of Western European thought, together with the growth of computing capability and transnational technological networks, led to the development of, what Jameson has called, the cultural logic of late capitalism. In this postmodern labyrinth, the old certainties of value and truth were shown to be nothing but illusion. Novels such as Thomas Pynchon's *Gravity's Rainbow* offered a literary way in which this new kind of ontology could be foregrounded, an ontology that had at its heart an awareness of its own fictionality. In terms of Kearney's schematic (1988), we had moved from the expressivist lamp to the labyrinthine hall of mirrors reflecting limitless variations on an ultimately illusory object.

It was from this maelstrom that hypertext fiction emerged as a new mode of storytelling. The term 'hypertext' itself was first used by the technologist, Theodor H. Nelson, in the early 1960s. Nelson had an idea for a new kind of text, a text that would explicitly embrace some of the affordances that computing technology now offered. The text would be 'hyper', a word that refers to something moving 'over' or 'beyond' established convention. What made it 'hyper' was that, in Nelson's vision, the text wouldn't exist as a single block of writing, but would be made up of 'text chunks' linked together and accessed through a computer screen (1993). Nelson used 'hypermedia' as an extension of his original term, in which a text was linked to graphics, audio and film, as well as plain text. Although Nelson was working hypothetically at this stage, he went on, with Andries van Dam, to develop the *Hypertext Editing System* at Brown University; the following year, in 1968, van Dam developed the *File Retrieval and Editing SyStem* (FRESS), a hypertext platform that could run on commercially available software. Independent of these projects, Douglas Engelbart, working at Stanford Research Institute, was able to demonstrate a hypertext interface in December 1968, at what has since come to be known as 'the mother of all demos' (Ryan 2010, 50).

The 1970s and 80s witnessed the increasing power and popularity of the personal computer (PC). Unlike the mainframe computer, PCs were all-in-one systems requiring little assembly or technical knowhow; and by the mid 1980s the computer had been brought out of the laboratory and could now be found in ordinary homes across the world. The significance of this is difficult to overestimate, although, amid all this activity, hypertext was pushed to the sidelines. Yet there was at least one important development: in 1987 Jay David Bolter and Michael Joyce released *Storyspace*, the world's first software platform designed specifically for the creation of hypertext fiction. The first story written for it was *afternoon, a story*. *Victory Garden* (1992) by Stuart Moulthrop and *Patchwork Girl* (1995) by Shelley Jackson followed on its heels.

It seems strange now to think of a time before the world wide web: a time before web pages and online media channels, social media platforms and mobile computing. Yet we don't have to turn the clocks back too far to reach such a place. Pre-web, there was the technical infrastructure of the internet, the global architecture of interconnected computer networks that developed from the late 1960s. ARPANET (Advanced Research Projects Agency Network), the precursor to the internet, slowly expanded throughout the 70s and 80s, linking military and research networks. Controlled by the American military, ARPANET was eventually decommissioned in 1990, opening up the global network to commercial use and the birth of the internet. By October 1993 more than two million computers were connected across it (Ryan 2010, 94).

This pre-web network activity was top down but there were also bottom up developments too. The deregulation of phone networks, in particular, was seized on by some as an unmissable opportunity to create user-led networks. XMODEM, the world's first modem for PCs, was released in 1977 by Ward Christensen. As Johnny Ryan notes, 'by the end of 1977 the phone system was open, inexpensive computer and modem equipment was commercially available and the software to allow them to communicate was in circulation free of charge' (2010, 68). In the following year, Christensen and his friends developed the *Computer Bulletin Board System* (bbs), the forerunner of web forums today. However, though significant, the bbs system was still extremely limited. The market leader remained CompuServe, the world's first online service provider. By the 1980s they provided subscribers with access to online content and messaging services; and by 1989 they also offered limited access to the internet (2010, 72). Yet, as late as the early 1990s, the internet was still a fragmented hotchpotch of services and content, very much in the shadow of ARPANET and its Cold War origins. Something needed to change.

In 1990 a researcher called Tim Berners-Lee at the European Organisation for Nuclear Research (CERN) set about trying to resolve some of the internet's issues. To do this, Berners-Lee returned to Nelson's original concept of hypertext. For Berners-Lee, hypertext offered a way of rethinking how content could be shared across the planet. It was this imaginative leap, the bringing together of two separate technologies, hypertext and the internet, into a single, unified, concept, that lay at the heart of the world wide web. As Berners-Lee writes in *Weaving the Web*, his own account of those years:

> Hypertext would be most powerful if it could conceivably point to absolutely anything. Every node, document – whatever it was called – would be fundamentally equivalent in some way. Each would have an address by which it could be referenced. They would all exist together in the same space – the information space.
>
> (1999, 16)

Through hypertext, the world wide web would be non-hierarchical and completely decentralised, where any content, regardless of format or type, could be freely and openly accessed from any other computer (1999, 15). Key to the realisation of Berners-Lee's vision was his development of *Hypertext Markup Language* (HTML). It was HTML that allowed the annotation and formatting of the individual text documents, or web pages. These pages were then accessed through an interface known as a web browser, which Berners-Lee also wrote.

The initial reaction to all of this was to pretty much ignore it. It wasn't until the release of the commercial web browsers, *Mosaic* in 1993, and *Netscape* in 1995, that the web really began to take off. In October 1994 there were eight million computers connected to the internet; by July 1997 this had increased to almost twenty million (Ryan 2010, 115). And with the rollout of Web 2.0 from 2004, websites were able to provide an enhanced range of functionality, including the ability to support user-generated content and interoperability, which in turn led to the development of social media sites such as Facebook (2004) and YouTube (2005).

By the mid 1990s, then, hypertext had been transformed from a niche format for digital content, to the fundamental architecture of the world wide web, a network that now has four billion users, over half the world's population (Kemp 2018).

If we are to understand the development of hypertext fiction in this period, it's critical we also understand this wider context. For some writers, hypertext fiction was an opportunity to engage directly with, what they saw as, a primary condition of contemporary life: its increasing mediation through digital networks. As I've already shown, the creative modalities of modernism and postmodernism can be seen as responses to the technological innovations of their time, and in particular, the enmeshment of the individual within globalised techno-capitalist infrastructures. The world wide web might have been born out of a philanthropic utilitarianism, but its foundations were the Cold War military architecture of ARPANET and the techno-scientific imperatives of nuclear research (CERN). Inscribed into the very essence of contemporary existence was the apparatus of political and social hegemony. The techno-military paranoia of *Gravity's Rainbow* had become manifest.

It is this 'wired world' that Sven Birkerts rails against in *The Gutenberg Elegies*, a book that first came out in 1994: 'as the world hurtles on towards its mysterious rendezvous, the old act of slowly reading a serious book becomes an elegiac exercise' (2006, 6). Yet, for others, this was exactly the reason to be thankful. Robert Coover led the way in this regard, with his *New York Times* article, 'The End of Books', published in June 1992. The print medium was 'a doomed and outdated technology'. And the fact that novels will soon be replaced by hypertext fiction should be warmly welcomed. For Coover, the novel was a 'virulent carrier of the patriarchal, colonial, canonical, proprietary, hierarchical and authoritarian values of the past' (1992, 11). Hypertext was nonlinear and interactive, and promised a new democratic age.

> With its webs of linked lexias, its networks of alternate routes (as opposed to print's fixed unidirectional page-turning) hypertext presents a radically divergent technology, interactive and polyvocal, favoring a plurality of discourses over definitive utterance and freeing the reader from domination by the author.
>
> (1992, 11)

There was therefore a rather uncomfortable duality about hypertext fiction. On the one hand, it emerged at the peak of postmodernism; as a form wrought from the very technologies that had helped engender postmodernist thinking, hypertext fiction seemed perfectly suited to reflect back this ontological imperative. As I've shown, this was an ontology born out of late capitalism, a condition characterised by deep scepticism and irony, in which truth exists only as an endless sequence of fictions.

Yet, at the same time, there was an undisguised optimism from some quarters about the possibilities that hypertext fiction now offered authors and their readers. One of the most important books here is George P. Landow's *Hypertext: The Convergence of Contemporary Critical Theory and Technology* (1992). As the subtitle suggests, Landow argued for the fundamental interconnection between hypertext and poststructuralist theory. For Landow, the inherent qualities of hypertext, it's openness and intertextuality, its empowering of the reader, provided what he saw as a kind of laboratory for the testing of poststructuralist theory (1992, 2). Drawing on Roland Barthes' concept of the 'readerly text' and Jacques Derrida's notion of the 'decentred book', Landow championed hypertext fiction as a vital new form of narrative construction: '[w]e find ourselves, for the first time in centuries, able to see the [printed] book as unnatural, as a near-miraculous technological innovation and not something intrinsically and inevitably human' (1992, 25).

He quotes Frederic Jameson quite early on, noting that 'everything is "in the last analysis" political' (Landow 1992, 30), tantalising us with a recognition of the postmodernist context within which poststructuralist thought sits. Yet this wider ontological condition quickly disappears, and instead Landow, like Coover, pursues a mainly benign vision in which the importance of hypertext is limited to the new affordances it offered authors: its ability to support multiple narrative or split-character perspectives, for example, or to foreground poetic or linguistic word play. As Coover noted, '[t]he most radical new element that comes to the fore in hypertext is the system of multidirectional and often labyrinthine linkages we are invited or obliged to create' (1992, 11).

Occasionally there are glimpses of the deeper cultural unease on which hypertext fiction sat. It comes in Landow's discussion of, what he terms, storyworlds (1992, 208), what we would today recognise as early single-player immersive game worlds, such as *Myst* (1993), *Freak Show* (1993) and *Bad Day at the Midway* (1995). Lander describes how, in these 'hypertext storyworlds', the reader becomes a form of postmodern detective, piecing together clues as they stumble through a fragmented story that deliberately disorientates and defamiliarises, frustrating their expectations of plot development and closure. This is the kind of detective story made famous by postmodern stories such as 'Death and the Compass' by Jorge Luis Borges (1942) and *The Name of the Rose* by Umberto Eco (1980). In this sense, hypertext storyworlds can be seen to offer a form in which a classic trope of postmodern fiction becomes a central feature of the narrative form (McHale 1992).

There were, of course, far-more avant-garde forms of experimentation within hypertext fiction than the storyworlds Landow describes, such as the works by Joyce, Moulthrop and Jackson that I've already introduced. These artists explicitly used hypertext as an ideal medium through which to explore the implications of poststructuralist theory on narrative form. Each of the authors offer multiple pathways through the story; the sequence of events is not fixed but dictated by the reader; there is no definitive ending but rather an endless series of possible outcomes; and the works deliberately undermine any sense of unity or narrative completeness. Whether or not intentional, the overall effect of these stories is one of profound disorientation and unease. The sort of questions these works pose to their readers are the sort of ontological questions that McHale associates with postmodern fiction: 'What is the mode of existence of a text, and what is the mode of existence of the world (or worlds) it projects? How is a projected world structured? And so on.' (1992, 10).

Mark Amerika's *Grammatron* (1997) is a hypertext fiction that falls squarely into this experimental category.[2] It is a very early example of a hypertext fiction written specifically to be hosted on the web. The story is essentially a retelling of the golem mythology in which a human-like creature is brought to life from clay or mud. In Amerika's version, Abe Golam, a successful digital artist, creates a sentient writing machine that he calls Grammatron. As the story itself states:

> GRAMMATRON uses the language of desire's own consciousness to disseminate the potential creative power that resides within the vast electrosphere. Meanwhile, the electrosphere as a whole is composed of an endless recombination of alphanumerical imagery that discharges itself as an evermorphing terrain of linguistic microbits threaded together in such a way so as to form the true signatures of all Digital Being.
>
> {signatures}

Like Jackson's *Patchwork Girl*, *Grammatron* is an adaptation of a creation myth. However, in the latter, the focus is very much the world wide web, with Abe Golam being a barely disguised Tim Berners-Lee, ultimately overwhelmed by what he has called into existence. As Amerika explained in an interview with Janelle Brown for *Wired*, *Grammatron* is ultimately about the search for meaning 'in a world dominated by information overload'. The story itself consists of more than a thousand text elements, thousands of links, many still and animated images, a background soundtrack and spoken audio (Rettberg 'American Hypertext' 2015, 31; see also Ensslin 2007, 104–106). *Grammatron* is therefore a synecdoche of the internet. By questioning the role and function of both the *Grammatron* and its creator, the reader is implicitly encouraged to think about the power structures operating through globalised information networks such as the web. Yet, ultimately, it is the interrelationship between language, text and consciousness that lies at the heart of the work.

Later hypertext fictions

The continuing rise in computing power, the growth of the web and technical developments such as Web 2.0, meant that the hypertext fictions of the 1990s quickly dated. Ensslin (2007) uses the terms second- and third- generation hypertext to indicate the movement from static web pages (first generation) to the use of multi (hyper) media (second generation). Third generation (cybertext) is characterised by what Ensslin describes as the 'transfer of control from user to machine, leaving the former increasingly powerless' (2007, 6).

Perhaps not surprisingly, reality is far messier, with the divide between first-, second- and third- generation hypertexts ultimately blurred and uncertain. A work such as *The Unknown* (1998), for example, sits uneasily between second and third generation definitions. Perhaps a better way of thinking about these changes is to position the debate, less from a technological perspective, and more from the point of view of the writers themselves. From this perspective, the debate becomes less of a discussion of technological teleology, and more about what writers, poets and authors actually wanted to do with the digital affordances presented to them. What becomes clear is that, as the new century progressed, the traditional notion of hypertext fiction began to fall out of fashion, at least at the cutting edge of avant-garde experimentation. Authors became less interested in the link-and-node affordances of hypertext *per se*. Instead, as I'll show in the following chapters, digital storytelling began to embrace user-led content; digital stories were released from the fixed screen of the PC, and writers began to experiment with spatiality through GPS-enabled smartphones. Rettberg calls these sorts of developments 'posthyperfiction', reflecting the fact that, as the first decade of the new century drew to a close, hypertext fiction, as an avant-garde art form, had indeed been left behind; but it also recognises that, in ways I'll be exploring in the following chapter, the lessons learnt by those early writers of hypertext fiction had not been completely forgotten.

Yet, even this is not the entire story. As it stands, Rettberg's notion of 'posthyperfiction' ignores the growth of hypertext fiction through grassroots, online communities, such as Twine and Genarrator. As Ensslin and Skains remind us, hypertext fiction might have become an irrelevance in avant-garde circles, yet elsewhere it underpins the growth of 'highly personalized and personalizable, autobiographical forms of hypertextual writing' (2018, 295). And we should not forget the impact of hypertextuality on the printed novel itself, in other words, the degree to which authors of physical books responded to the pervasive affordance of the web. Understanding both these phenomena are a key part of understanding the postdigital poetics of hypertext fiction.

By the time the first iPhone was released in 2007, the classic hypertext fiction as an experimental art form was already dead. *The Doll Games* (2001) by Shelley and Pamela Jackson is one of the last examples of the classic 'link and node' hypertext fictions that had first started to appear in the mid 1980s.[3] While Jackson's *Patchwork Girl* (1995) had been written using the Storyspace software and distributed on CD-ROM, *The Doll Games*, like *Grammatron* four years earlier, was published straight onto the web. Rather like *The Unknown* (1998), *The Doll Games* explores a collaborative/user-generated approach to digital storytelling, and as such, sits right on the cusp of the Web 2.0 revolution.

Like *The Unknown* too, *The Doll Games* provides an ironic and playful exploration of postmodernity. It is a fragmented work, consisting of numerous documents and textual forms, such as interview transcripts, photographs and other artefacts (including descriptions of twenty dolls and a glossary). The material appears to have been put together by an academic, J. F. Bellwether, PhD; it is the fictional Bellwether who provides a meta-narrative in *The Doll Games*, commenting on the reception of the very work in which he appears:

> Giving the Postmodern pastiche a comradely nod but eschewing its cynicism, the Doll Games have confused some critics. Never afraid of acknowledging wish-fulfilment as narrative's *primum mobile,* the Doll Games presented a resolutely cheerful *Weltanshauung,* leading some scholars to dismiss them as naive.
>
> {definition}

As I've already said, what makes *The Doll Games* particularly interesting is its nod towards user-generated writing. On the content page, the last section is invitingly entitled 'Want to Play?'. When clicked, the reader is taken through to a page in which Shelley and Pamela directly ask readers to share their own personal experiences of doll games.

> We are looking for true stories. We know that lies are sometimes more interesting, and we may not be able to tell if you are lying. However, we are telling the truth, and we think that makes our project better. On the other hand, if you send us fiction and we like it, we might make a special section just for that– you never know.
>
> {question}

At least two user-generated stories have been added to a section of the website called 'Interviews'. The rest of the material appears to be transcripts of short interviews in which various people talk to Shelley about their experience of doll games.

Looked at today, *The Doll Games* feels rather curious. It has the aesthetics and structure of web 1.0 – there's text to be read, and links to be clicked, but in terms of direct interaction, that's pretty much it. Yet the request for user-generated content to be emailed to the authors hints at a growing interest in user-generated content that would soon be mainstreamed with the rollout of Web 2.0. *The Unknown* is the same. It's worth noting that user-generated content goes back to the very beginnings of hypertext fiction. Michael Joyce's *afternoon: a story* (1987), for example, allowed users to write responses into text boxes. Yet, here the text boxes were very much an add-on, structurally separate from the story itself. This is not the case for either *The Unknown* or *The Doll Games* which are far closer to Joyce's own definition of constructive hypertext. Both represent a creative endpoint of Web 1.0, a rather clunky attempt at user participation that would soon be swept away by Web 2.0 and the social media revolution.

Yet there's something else too. Both *The Unknown* and *The Doll Games* have many of the characteristics of classic postmodern texts. Like *The Unknown*, *The Doll Games* is fragmented, ironic and without any clear narrative structure or pathway. In the same way that J. F. Bellwether, PhD, was used to satirise cultural hierarchies, the very aesthetics and form of *The Doll Games* parodied the world wide web as an enclosed patriarchal knowledge system. As an active part of that very knowledge system, *The Doll Games* should be seen as a kind of performative intervention, in which the ideological and gendered workings of the web were foregrounded as 'a materially distinct writing platform that itself affects and shapes work developed there' (Rettberg, 'Posthyperfiction' 2015, 213). And yet, the participatory elements of *The Unknown* and *The Doll Games*, also speak of something else. In both cases – the collaborative writing of *The Unknown* and the user-generated content of *The Doll Games* – the participatory elements undermine the overt postmodernity of the work. This is particularly marked in *The Doll Games*. It's as though there are two stories – the classic labyrinthine-structured postmodern hypertext, with its denial of narrative form and its deep cynicism of any kind of knowledge or meaning; and a far-more intimate, 'real', story with its deep desire to connect with the wider world by 'eschewing … cynicism'. In the first story, Shelley and Pamela appear as ironic avatars, constructs in a story where nothing is what it seems. In the second, the real authors suddenly break cover, reaching out to us as fellow human beings. The user-generated stories are an invitation to 'come play', to emotionally connect with Shelley and her sister, and through that playful interaction, to make sense of the world.

Understanding this through a technological taxonomy of first-, second- or third-generation hypertexts, misses an important point. These two hypertext fictions are not just at the cusp of a *technological* revolution (Web 2.0); I would also argue that they sit on the transition between two *creative* modalities: postmodernity and, what I have termed, map/rhizome/string-figure. In terms of our schematic, the stories are fundamentally Janus-faced, looking back to the postmodernity of the labyrinth, at the same time that they also implicitly respond to a new kind of participatory connectivity. Although odd and awkward to our eyes, both *The Unknown* and *The Doll Games* can be considered to be some of the earliest forays into this new form of creative modality. Despite being redolent with postmodern characteristics, inscribed into both stories is a new kind of imperative: the desire for an affective experience, embracing both a situated and embodied understanding of storifying.

The story does not end here of course. As we'll see in the next chapter, digital authors began to produce work that was far-more complex and hypermedial in nature. Functionally and aesthetically these twenty-first-century stories are very different experiences from that of the humble hypertext as it was first conceived. Yet it makes sense to pause here and take stock before we move on.

Summary

What I've tried to do in this chapter is situate the development of postdigital storytelling within an historical continuum. Specifically, I've explored the development of hypertextual storytelling from its earliest conceptual beginnings in the late 1960s through to the turn of the new century. By doing so what I've been able to foreground is the way these hypertext fictions were informed by the shifting creative modality from labyrinth to map/rhizome/string-figure, a transformation that was itself a product of the waning of postmodernism.

By extending my analysis to the early nineteenth century I've also been able to explore the ongoing relationship between storytelling as a mode of representation and the social and cultural effects of technological change. Danius' 'crisis of the senses' (2002, 3) and Trotter's meta-attitude (2013, 273) are different ways of describing the fundamental shock of the new, a revolution of the mind in the way people interacted with the world. For writers experiencing these drastic and sudden changes, the imperative was to find a form of words that could somehow respond to such transformations. Just as James Joyce and Virginia Woolf strove to find a new way of representing the deep-rooted epistemological changes of the early twentieth century, so too did Michael Joyce and Shelley Jackson reflect an avant-garde of digital artists, desperate to explore the poststructuralist sublime of the hypertextual narrative. As we've seen, these stories were fragmented and polyvocal, deliberately disorientating and defamiliarising, frustrating the reader's expectations of plot development and closure. Landow stated in 2006 that most hypertext authors write from the dawn of the end of print (*Hypertext 3.0* 2006, 135), a sort of poststructuralists' nirvana. And yet early millennial stories such as *The Unknown* and *The Doll Games* indicate that if there is indeed a new dawn, then it is less about the end of print *per se*, and much more to do with the end of postmodernism all together and the rise of a completely new way of seeing the world. In works such as *The Unknown* and *The Doll Games*, writers were slowly emerging from the labyrinth, blinking in the light of, what I and others have termed, a metamodernist dawn. Understanding in more detail this crucial stage in the evolution of digital storytelling becomes the focus of the next chapter.

Notes

1 See http://unknownhypertext.com.
2 See www.grammatron.com.
3 See www.ineradicablestain.com/dollgames.

Works cited

Amerika, M. *Grammatron*, 1997. Web. 5 February 2018. www.grammatron.com.
Barthes, R. 'The Death of the Author'. *Aspen* 5–6, 1967, np.

Baudrillard, J. *Simulations*. Trans. Beitchman, P. et al. New York: Semiotext(e), 1983.
Benjamin, W. 'The Author as Producer'. *Selected Writings*. Eds, Eiland, H., Jennings, M. W. and Smith, G., Trans. Jephcott, E., Vol. 2. Cambridge, Massachusetts: Belknap, 1996, 768–782.
Benjamin, W. 'The Work of Art in the Age of its Technological Reproducibility: Second Version'. *Selected Writings*. Eds, Eiland, H. and Jennings, M. W., Trans. Jephcott, E. and Zohn, H., Vol. 3. Cambridge, Massachusetts: Belknap, 2002, 101–133.
Berners-Lee, T. *Weaving the Web: The Original Design and Ultimate Destiny of the World Wide Web by its Inventor*. London: Harper, 1999.
Birkerts, S. *The Gutenberg Elegies: The Fate of Reading in an Electronic Age*. New York: Farrar, Straus and Giroux, 2006.
Borges, J. L. 'Death and the Compass'. *Labyrinths: Selected Stories and Other Writings*. Eds, Yates, D. A. and Irby, J. E. London: Penguin, 2000, 106–117.
Brown, J. 'Amerika's Fragmented Pages'. *Wired*. 27 June 1997. Web. 26 May 2018. www.wired.com/1997/06/amerikas-fragmented-pages.
Coover, R. 'The End of Books'. *New York Times*, 21 June 1992, 11.
Derrida, J. *Writing and Difference*. Chicago: University of Chicago Press, 1978.
Eco, E. *The Name of the Rose*. Milan: Bompiani, 1980.
Ensslin, A. *Canonizing Hypertext: Explorations and Constructions*. London: Bloomsbury, 2007.
Ensslin, A. *Literary Gaming*. Cambridge, Massachusetts: Massachusetts Institute of Technology, 2014.
Ensslin, A. and Skains, L. 'Hypertext: Storyspace to Twine'. *The Bloomsbury Handbook of Electronic Literature*. Ed, Tabbi, J. London: Bloomsbury, 2018, 295–309.
Danius, S. *The Senses of Modernism: Technology, Perception and Aesthetics*. Ithaca, New York: Cornell University Press, 2002.
Douglas, J. Y. ' "How Do I Stop This Thing?" Closure and Indeterminacy in Interactive Narratives'. *Hyper/Text/Theory*. Eds, Landow, G. P. Baltimore: John Hopkins University, 161–188.
Gillespie, W., Rettberg, S., Stratton, D. and Marquardt, F. *The Unknown*. 1998. Web. 30 April 2018. http://unknownhypertext.com.
Goble, M. *Beautiful Circuits: Modernism and the Mediated Life*. New York: Columbia University Press, 2010.
Harvey, D. *The Condition of Postmodernity*. Oxford: Blackwell Publishing, 1990.
Hayles, N. K. *Electronic Literature: New Horizons for the Literary*. Notre Dame, Indiana: University of Notre Dame Press, 2008.
Herman, L. and Weisenburger, S. Gravity's Rainbow, Domination, and Freedom. Athens, Georgia: University of Georgia Press, 2013.
Heslop, H. *Last Cage Down*. London: Wishart, 1935.
Hunter, A. *The Cambridge Introduction to the Short Story in English*. Cambridge: Cambridge University Press, 2007.
Jackson, S. *Patchwork Girl; or a Modern Monster by Mary/Shelley and Herself*. Watertown, MA: Eastgate Systems, 1995.
Jackson, S. and Jackson, P. *The Doll Games*, 2001. Web. 28 May 2018. www.ineradicablestain.com/dollgames.
Jameson, F. *Postmodernism or, The Cultural Logic of Late Capitalism*. London: Verso, 1991.
Joyce, M. *afternoon: a story*. Watertown, MA: Eastgate Systems, 1987.

Joyce, M. 'Hypertext and Hypermedia'. *Of Two Minds: Hypertext, Pedagogy and Poetics*. Ann Arbor: University of Michigan Press, 1995 [1993], 19–30.

Joyce, M. *Of Two Minds: Hypertext, Pedagogy and Poetics*. Ann Arbor: University of Michigan Press, 1995.

Joyce, M. 'Siren Shapes: Exploratory and Constructive Hypertexts'. *Of Two Minds: Hypertext, Pedagogy and Poetics*. Ann Arbor: University of Michigan Press, 1995 [1993], 39–59.

Kacandes, I. *Talk Fiction: Literature and the Talk Explosion*. Lincoln: University of Nebraska Press, 2001.

Kearney, R. *The Wake of Imagination*. London: Taylor and Francis, 1988.

Kemp. S. *Digital in 2018: Essential Insights into Internet, Social Media, Mobile, and Ecommerce use Around the World*. London: We Are Social/Hootsuite, 2018.

Landow, G. P. *Hypertext: The Convergence of Contemporary Critical Theory and Technology*. Baltimore: John Hopkins University, 1992.

Landow, G. P. *Hypertext 3.0: Critical Theory and New Media in an Era of Globalization*. Baltimore: John Hopkins University, 2006.

Landow, G. P., (ed.) *Hyper/Text/Theory*. Baltimore: John Hopkins University Press, 1994.

Lodge, D. *After Bakhtin: Essays on Fiction and Criticism*. London: Routledge, 1990.

McHale, B. *Constructing Postmodernism*. London: Routledge, 1992.

McHale, B. *Postmodernist Fiction*. London: Routledge, 1999.

Menke, R. *Telegraphic Realism: Victorian Fiction and Other Information Systems*. Stanford, California: Stanford University Press, 2008.

Moulthrop, S. *Victory Garden*. Watertown, MA: Eastgate Systems, 1992.

Nelson, T. *Literary Machines*. Sausalito, California: Mindful Press, 1993.

Page, R. and Thomas, B. 'Introduction'. *New Narratives: Stories and Storytelling in the Digital Age*. Eds, Page, R. and Thomas, B. Lincoln: University of Nebraska, 2011, 1–16.

Pynchon. T. *Gravity's Rainbow*. New York: Viking Press, 1973.

Rettberg, S. 'All Together Now: Hypertext, Collective Narratives, and Online Collective Knowledge Communities'. *New Narratives: Stories and Storytelling in the Digital Age*. Eds, Page, R. and Thomas, B. Lincoln: University of Nebraska, 2011, 187–204.

Rettberg, S. 'The American Hypertext Novel, and Whatever Became of it?' *Interactive Digital Narrative: History, Theory and Practice*. Eds, Koenitz, H., Ferri, G., Haahr, M., Sezen, D. and Sezen, T. I. London: Routledge, 2015, 24–35.

Rettberg, S. 'Posthyperfiction: Practices in Digital Textuality'. *Interactive Digital Narrative: History, Theory and Practice*. Eds, Koenitz, H., Ferri, G., Haahr, M., Sezen, D. and Sezen, T. I. London: Routledge, 2015, 174–184.

Ryan, J. *A History of the Internet and the Digital Future*. London: Reaktion Books, 2010.

Thomas, B. *Fictional Dialogue: Speech and Conversation in the Modern and Postmodern Novel*. Lincoln: University of Nebraska Press, 2012.

Trotter, D. *Literature in the First Media Age: Britain Between the Wars*. Cambridge, Massachusetts: Harvard University Press, 2013.

Uricchio, W. 'Historicising Media in Transition'. *Rethinking Media Change: The Aesthetics of Transition*. Ed, Thorburn, D. and Jenkins, H. Cambridge, Massachusetts: MIT Press, 2003, 23–38.

Uzanne, O. 'The End of Books'. *Scribner's Magazine*, 16, 1894, 221–231.

van den Akker, R. Gibbons, A. and Vermeulen, T., eds. *Metamodernism: Historicity, Affect, and Depth After Postmodernism.* London: Rowman & Littlefield International, 2017.

Waugh, E. *Vile Bodies*. London: Chapman & Hall, 1930.

Woolf, V. 'The Mark on the Wall' [1919]. *The Mark on the Wall and Other Short Fiction*. Oxford: Oxford University Press, 2008, 3–10.

Woolf, V. 'Poetry, Fiction, and the Future'. *The Essays of Virginia Woolf*. Ed, McNellie, A. Vol. 4, New York: Harcourt Brace Jovanovich, 1988, 433–434.

5 Postdigital hypertextuality

Introduction

The Virtual Disappearance of Miriam (2000) by Martyn Bedford and Andy Campbell is a digital story from the turn of the millennium.[1] At one level it operates as a conventional postmodern story, both in terms of the narrative but also the structural design of its emplotment. Although the story is about a character called Luther who is searching for his girlfriend, Miriam, after her unexpected disappearance, it quickly takes a metafictional direction. This is particularly overt in the section 'Playing the Male Lead' in which Luther seems to step outside the fictionality of the story to engage in conversation with Hollywood stars, Harvey Keitel and Quentin Tarantino. As Luther himself recognises within the story, he has entered a 'Labyrinth of the Bizarre'. This postmodern sensibility is emphasised through the structure of the story too. The home page offers the reader a choice of four sections: Missing You Already; House of Sam; Playing the Male Lead; and Miriam. The choice the reader makes has a fundamental impact on how they experience the story.

So far, so postmodern. And yet *The Virtual Disappearance of Miriam* is not quite as postmodern in its intent as it at first appears. Take those four sections, for example, shown in Figure 5.1. They are rendered sequentially across the screen and are clearly numbered. As James Pope notes in his own analysis of the story, 'even though the author offers a menu of chapters at the start, it is clear what the "conventional" order would be' (2006, 461), the implicit invitation being to start with chapter one, and move through numerically. This reaching out for narrative structure and coherency is then repeated at the very end of the story. As Figure 5.2 illustrates, the reader is presented with three options, the happy, sad or postmodern ending. The first two offer what I would call traditional endings. In the postmodern ending, however, it's suggested that Miriam never really existed in the first place.

I would argue that *The Virtual Disappearance of Miriam* offers the kind of transitional story of the sort we've already seen with *The Unknown* (1998) and *The Doll Games* (2001). All three were made around the turn of the century and all three offer the first signs of an escape from the postmodern labyrinth.

Figure 5.1 *The Virtual Disappearance of Miriam* (2000). Screenshot.
Source: © Martyn Bedford and Andy Campbell (originally commissioned and published by the Ilkley Literature Festival in conjunction with Route Publishing).

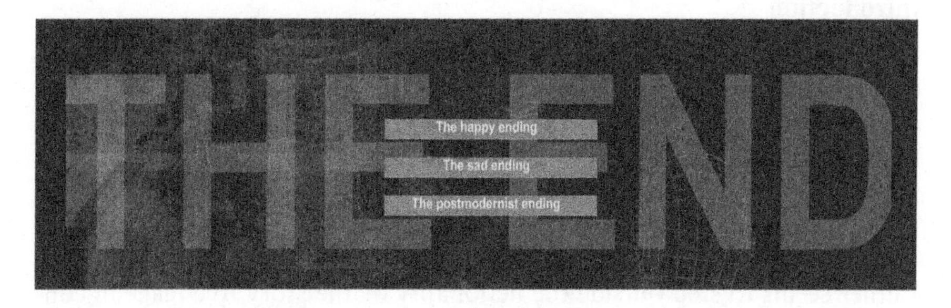

Figure 5.2 *The Virtual Disappearance of Miriam* (2000). Screenshot.
Source: © Martyn Bedford and Andy Campbell (originally commissioned and published by the Ilkley Literature Festival in conjunction with Route Publishing).

With *The Virtual Disappearance of Miriam*, for example, it's as though we have two stories, a more straightforwardly-affective one, looking forwards into the future, and a more complex, postmodern one, looking back to the influences of the previous two decades. This moment of hesitancy, this authorial polyvalence, is clear evidence of that gradual transition in creative modality from postmodernism to metamodernism that has been, and will remain, the context of this book.

Both this and the previous chapter stand either side of this pivot. Whereas Chapter 4 explored the evolution of hypertext stories from their beginnings, this chapter will discuss what happened next, bringing their evolution to the present day. As David Ciccoricco notes, the 'form and function of links and nodes' marked the emergence of a 'distinctly new narrative poetics' (2012, 479). Importantly, the chapter will foreground the transition towards postdigitality within these poetics, situating this transformation firmly within wider ontological change. Not only will the chapter examine key digital works, but it will also analyse the impact of hypertextuality on the printed form, charting the development of increasingly complex hybridic entanglements across the digital and the non-digital. As Scott Rettberg extolls, '[h]ypertext is dead. Long live hypertext' ('Posthyperfiction' 2015, 183).

Hypertextuality: the new millennium

Chapter 4 was focussed on the classic period of hypertext fiction, a period of perhaps no more than fifteen years, that ran from the mid 1980s. As I've made clear, this was a period very much under the influence of postmodern thinking, and many contemporaneous hypertext fictions can be seen as explorations of this condition. Here, the digital and the non-digital were seen as two mutually-exclusive domains: the non-digital represented the old analogue world that would soon be gone; the digital, in comparison, represented a new era, in which the political and social hierarchies of the past would be demolished, and in its place would evolve a new relativised network of knowledge and knowledge construction. Mediality was therefore perceived as a zero sum game, in which, as the digital domain expanded, the non-digital would naturally wither away. For early adopters such as Landow, Joyce, Bolter and Coover, hypertext fiction was seen as part of this technological vanguard that would naturally consign print-based media to the historical dustbin. Bolter's 'late age of print' had truly arrived (2001, 3).

The theoretical underpinnings of postmodernism naturally reinforced such a binary model. If late capitalism marked an ontological juncture in human history, then it helped to underline the degree to which this rested on a stepwise change in technological innovation. As we've seen, hypertext fiction such as Joyce's *afternoon: a story* (1987) and Moulthrop's *Victory Garden* (1992), trod a delicate line by embracing both the deeply ironic scepticism of postmodern philosophy, and the 'empowering' and 'liberating' potential of one of its key technological underpinnings, namely hypertext and the web.

And yet it would be wrong to suppose that the condition of hypertextuality didn't also have an effect on print-based storytelling. After all, we've already seen how technology influenced the form and content of the modernist and postmodernist novel. This phenomenon didn't stop with *Gravity's Rainbow* (1973). Although a sustained study of this remains outside the parameters of this book, it is important that some recognition of the impact of hypertextuality on the printed book is nevertheless made. As I explained in Chapter 3, we need to be able to think *across* as much as *within* our domains.

It is important because what I'll be arguing here is that the nucleus of postdigitality can be found in the way *both* printed and hypertext fiction from the late 1990s were increasingly entangled in their representations of and responses to the hypertextual condition. This reaching out operates from both domains, from the digital to the non-digital, but also vice versa. As Hayles notes: 'digital technologies do more than mark the surface of contemporary print novels. They also put into play dynamics that interrogate and reconfigure the relations between authors, and readers, humans and intelligent machines, code and language' (2008, 186). Hypertext authors such as Michael Joyce (1995) and Jay David Bolter (2001) have written openly about how their own work drew on and then transformed the characteristics of the physical book. It should therefore not be a surprise that the reverse happened too.

The printed fiction I want to focus on is *House of Leaves* by Mark Z. Danielewski (2000). The novel quickly became a publishing sensation and the book continues to be a bestseller, having reached canonical status as a postmodern horror story. The novel consists of a complex interweaving of nested stories including the first-person narrative of Johnny Truant, Zampanò's manuscript and the transcripts and notes of various interviews. The novel has at its heart the story of the Navidson family who, on returning from a holiday, find that a strange door has suddenly appeared in their home. Over the coming weeks, other spatial anomalies take place which eventually lead into an endless series of rooms, passageways and stairs. The father, Will Navidson, films and photographs as he explores these phenomena, the resulting film being known as *The Navidson Record*.

As Katharine Cox notes, the labyrinth at the heart of *The Navidson Record* is mirrored by the labyrinthine structure of the novel itself. *The Navidson Record*, as a documentary film, remains outside *The House of Leaves*: instead, the reader accesses it indirectly through the fragmentary Zampanò manuscript, which itself is revealed through the narration of Johnny Truant. Unnamed editors of Truant's text, evidenced in their footnoting, forms a further narrative level to the story.

As we've seen, *Gravity's Rainbow* is a sprawling, fragmentary novel that is also seen as classically postmodern. And yet *House of Leaves* and *Gravity's Rainbow* are very different works. *Gravity's Rainbow* is ontologically and structurally anarchic; the work itself seems determined to undermine the very notion of its own ability to make sense of reality. Parody, satire and irony are key features. In comparison, *House of Leaves* comes from a completely different ontological position. The irony, parody and satire are gone. While *Gravity's Rainbow* delights in a textual anarchy freed from imposed ethical and moral constraints, *House of Leaves* positions itself as an earnest exploration of the relationship between knowledge and reality. The latter has much more in common with modernism in that sense: at its heart, *House of Leaves* is an investigation into the human condition at a time of significant technological change. As Alison Gibbons states in her own analysis of Danielewski's novel (2012), *House of Leaves* is not postmodern at all, but rather representative of a paradigm shift 'in the way we see the world and our place within it' (2012, 3). In other words, *House of Leaves* is an example of what Gibbons, amongst others, have labelled metamodern.

While Gibbons sees *House of Leaves* as a classic 'multimodal' printed novel, in other words, one that uses 'multiple semiotic modalities, primarily the verbal and visual' (2012, 1), Hayles is surely right to interpret this as a direct response to changes within the contemporary technological landscape. As Hayles states:

> In *House of Leaves*, the recursive dynamic between strategies that imitate electronic text and those that intensify the specificities of print reaches an apotheosis, producing complexities so entangled with digital technologies

> that it is difficult to say which medium is more important in producing the novel's effects.
>
> (2008, 175)

Hayles describes how *House of Leaves* manifests four characteristics of a digital text: it is multimodal, layered (with overwrites and gaps), performative and temporally fractured (2008, 163–165). For Hayles, digital technology runs through every facet of the book, from its explicit referencing of film, video, photography, telegraphy and graphics, to the underlying code of its representation. Gibbons describes how the layout of the various texts encourages an active interaction with the novel, a physicality in which the book itself is rotated and turned at the same time as the reader mentally explores the narrative: '[t]he reader of *House of Leaves* does not just "read" the novel. Like Navidson, and aligned with him though subjective and corporeal resonance, s/he seems to actively explore it' (2012, 81).

For both Hayles and Gibbons, then, it is form and structure that is key to understanding the influence of technology on *House of Leaves*, from its recursive sequence of nested narratives through to its pronounced multimodality. Yet I would argue their analyses fall short in fully appreciating the extent to which *House of Leaves* was influenced by hypertextuality, in particular. The novel itself was written during the 1990s, at the moment when the world wide web was becoming a global phenomenon. Early versions of the novel were first published on the internet and elicited a cult following through online discussion boards (Pressman 2006, 119; Thomas 2011). Danielewski went on to inscribe this print-to-web-to-print publication history within the novel itself: the footnotes written by the anonymous editors indicate that versions of Johnny Truant and Zampano's manuscript had been previously uploaded to the internet. As a result of this, a number of emails were received by the editors, one of which is given in full: 'I think Johnny was a little off here. I wanted to write and tell you about it' (*House of Leaves* 2000, 151). Further on in the novel, another footnote from the editors reads, 'Note: This section also elicited several e-mails' (*House of Leaves* 2000, 263). Emails from Clarissa, Natalie and Bethami then follow: 'Lude was such a jerk and a shitty fuck. You can tell him that – Clarissa April 13, 1999' (*House of Leaves* 2000, 263). The overall effect of these fictional annotations is to deliberately evoke the processes and formulations of online publishing, a phenomenon that was very much still in its infancy in the late 1990s. A text is uploaded to the internet; a readership is allowed to comment and discuss the text; and then finally, the outcome of those discussions finds its way back into the text. It was exactly this kind of positive circularity that lay at the heart of the web which, for Berners-Lee, was always first and foremost a 'social creation … to help people work together' (1999, 133). Jessica Pressman calls *House of Leaves* a 'networked novel' and it's very easy to understand why. The computational aesthetics of the web permeate the entire novel, both in its physical design but also the playful fictionality of the wider productive milieu from which the novel emerged.

Within the book itself, the nested narrative architecture is linked through the annotations. But the specific colouring of the word 'house' within the entirety of the text also appears to replicate the aesthetic of a hyperlink. These hyperlinks work across the novel but also, potentially, reach outwards, beyond the text itself. In fact this sense that the novel exists within a wider ecosystem of paratexts is one deliberately fostered by Danielewski. As Pressman notes, '*House of Leaves* is the central node in a network of multimedia, multiauthored forms that collectively comprise its narrative' (2006, 107). This external network consists of a variety of non-digital content published more or less concurrently with the novel, including *The Whalestoe Letters* (2000), a series of letters between Johnny Truant and his mother, and the music album, *Haunted* (2000), by the recording artist Poe, Danielewski's sister. The relationship between *House of Leaves* and *Haunted* is particularly interesting; although both exist as independent works, the printed book and the music are indelibly intertwined. Tracks on the latter, for example, include 'Dear Johnny', '5&½ Minute Hallway' and 'House of Leaves'. Early editions of *House of Leaves* explicitly advertised the album on its back cover; the quote 'No one should brave the underworld alone' (423) is mischievously attributed to 'Poe' in the novel. The assumption of course is that the author is Edgar Allan Poe when in fact it is the author's sister.

The use of paratexts is nothing new. J.R.R. Tolkien famously generated a rich tapestry of paratexts related to Middle Earth over many years. Yet the paratextual nature of *House of Leaves* feels very different: the contemporaneity of both the *Whalestoe Letters* and *Haunted*, the embedded links between them, indicate a desire to create a new kind of novel, one in which the nested interconnectivity of *House of Leaves* is extended beyond the physicality of the printed book. As Pressman notes, these paratexts deliberately seek to re-envision the printed work as a node within a wider multimedia network. And the aesthetics of hypertextuality are central to this. As well as being inscribed with Poe's web page address, the novel also boasted its own web address (www.houseofleaves.com), one of the first novels to do so. The website is still active although it looks very different from how it would have appeared when the novel was first published. Using the Internet Archive (The WayBack Machine) it is possible to access a snapshot of the website from 28th November 2001. At that time, any visitor to the home page was automatically transferred to a discussion board page, shown in Figure 5.3. There's a number of things to say about this page. First, the colouration of the word 'house' in the novel is continued through onto the website. This imitates a network of nodal hyperlinks that stretches across both the printed and digital domains. Second, the relationship between *House of Leaves*, *The Whalestoe Letters* and *Haunted* is made explicit on the website with the three distinct, but interrelated, discussion boards. And third, those discussion boards self-consciously mirror the (fictionalised) online-communities through which

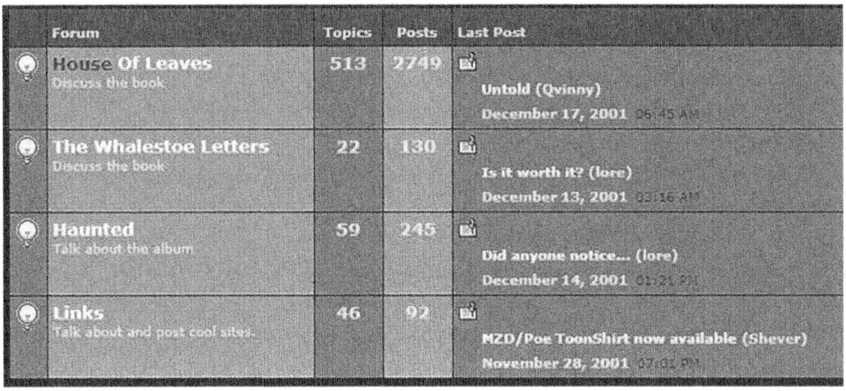

Forum	Topics	Posts	Last Post
House Of Leaves Discuss the book	513	2749	Untold (Qvinny) December 17, 2001 06:45 AM
The Whalestoe Letters Discuss the book	22	130	Is it worth it? (lore) December 13, 2001 03:16 AM
Haunted Talk about the album	59	245	Did anyone notice... (lore) December 14, 2001 01:21 PM
Links Talk about and post cool sites.	46	92	MZD/Poe ToonShirt now available (Shever) November 28, 2001 07:01 PM

Figure 5.3 *House of Leaves* website. Screenshot of Discussion Boards page. 28 November 2001 (Internet Archive, Wayback Machine).

Source: © Mark Z. Danielewski.

House of Leaves supposedly evolved. In this way, the online boards offered the novel's readers a way to become part of the story itself. As the editors make clear in their foreword to the novel, *House of Leaves* is a work in progress which they will gladly revise and correct in subsequent editions (*House of Leaves* 2000, vii).

The overall effect then is of a 'networked novel', just as Pressman describes; in other words, *House of Leaves* is a novel that explicitly engages with knowledge construction in the hypertextual and multimedia context of the world wide web. As Mark C. Taylor explains, *House of Leaves* 'is a web, and the web is a house of leaves' (2013, 139).

The implications of all this are profound: rather than understanding the novel as a digital text, as Hayles describes, I would argue that *House of Leaves* is actually a very early example of *postdigitality*. Rather than endorsing a binary model of digital/non-digital domains, *House of Leaves* explicitly seeks to present a hybridised ecosystem, in which the story evolves across a transmedial landscape, including physical print, music and digital text. One of the biggest clues that indicates the importance of this transmediality to the novel is the fact that Danielewski has not allowed *House of Leaves* to appear in electronic form. There is no ereader version of the novel. Speculation about this amongst the fanbase often highlights the impossibility of digitising the novel's typography. Yet this does not seem like an adequate enough reason: there have been lots of experimental, multimodal novels that have been successfully digitised. Ultimately, the only reason that would stop the digitisation of *House of Leaves* would be the withholding of permission by the author. Although Danielewski has not spoken openly about this, it does not seem farfetched to assume that his justification for this is one very much influenced by his creative vision for the work. It is certainly not the reaction of a technophobe – as we've seen, Danielewski was one of the first authors to directly engage with the web as part of his practice. Instead, the decision suggests a creative response in which the specificities of media are understood by the author to be a fundamental aspect of the overall work. In other words, the transmediality of *House of Leaves*, and, in particular, the enmeshment of the story across print-based and digital platforms, is a critical part of the narrative. The reader turns the printed book around in their hands as they attempt to follow the typography; but then they switch to their computer to access the discussion boards and to play the CD of Poe's *Haunted.* I would argue that *House of Leaves* was published with this transmedial interactivity encoded into its design from the very beginning. As Taylor notes in his own analysis of *House of Leaves*: 'we have been online all along, even when we thought we were unplugged…' (2013, 155).

Like Gibbons (2012), Mary. K. Holland also sees *House of Leaves* as indicative of a new kind of creative modality. For Holland, the novel has all the hallmarks of a classic postmodern text. Yet it has something else too.

> But ultimately it is not the novel's insistent postmodernity, its defiance of conventional novelistic devices, narrative construction, and subject matter that make this book remarkable at the "end" of the postmodern age, but rather its surprisingly conventional appeal to exactly the emotion and psychological depth that much of postmodernity had long jettisoned.
> (2013, 98)

While *House of Leaves* has all the outward characteristics of a postmodern text, it is not actually a postmodern novel. It is something else. For both Holland and Gibbons, what makes *House of Leaves* different from the classic postmodern texts that preceded it is its exploration of authentic emotional affect. Fundamentally, the novel is about familial reconciliation (Holland 2013), yet beyond that Taylor also sees a restless search for meaning and psychological connection. For Holland, it is this emphasis on 'affect, earnestness, subjectivity, and family' (2013, 99) that makes *House of Leaves* an early product of a new creative and cultural modality, one in which irony and depthlessness is replaced by a renewed imperative for writing that elicits 'empathy, communal bonds, ethical and political questions, and, most basically, communicable meaning' (2013, 17). Like Gibbons, Holland uses the term 'metamodernism' to describe this new period, re-emphasising the modernist ideals of self and meaning, but resituating that within poststructuralist concepts of language and arbitrariness. It is *House of Leaves'* embracing of this new ontological position that makes it such an innovative and important work.

Although neither Gibbons nor Holland specifically discuss the technological aspects of the novel, I would argue that the calculated enmeshment of *House of Leaves* across digital and non-digital domains is also a fundamental aspect of this new creative paradigm, what I have labelled map/rhizome/

string-figure. In other words, what makes *House of Leaves* a metamodernist work is not just its embracing of a new kind of psychological realism. Critical here is also how the novel explored the way human subjectivity and knowledge structures at the turn of the new century were increasingly transmedial in nature. It is this hybridised performance that makes *House of Leaves* a very early example of a postdigital novel, one that was reaching out across the non-digital domain into the digital. Of course, at this point in time, it was the non-digital domain that was still dominant; the digital paratext were subordinate to the physical book. This hybridisation was not only expressed through the physical interconnection between the printed novel and the web, but hypertextuality was also inscribed onto the very form and typology of the printed page itself.

By the beginning of the new century, two important developments were taking place in regards to digital storytelling. The first was that hypertext stories were slowly evolving from the postmodern, labyrinthine structure that lay at their roots. As we've seen, *The Unknown* (1998) experimented with collaborative composition across a delimited group of fellow artists.[2] The primary output was a sprawling hypertext novel; yet, in its effort to parody an authors' book tour, it eventually came to include real-life art projects, live readings and a press kit. As one of its authors notes: '*The Unknown* was thus both a novel and a work of performance writing and in some respects also a constraint-driven writing game' (Rettberg 'American Hypertext Novel' 2015, 32). Shelley and Pamela Jackson's *The Doll Games* (2001) is another hypertext story built around a heteroglossic structure.[3] In this case, however, the work included an open invitation for readers to submit their own material for inclusion in the story. As the authors themselves implore: 'If we include your contribution, we will give you credit and post a brief biography, so please tell us how old you are, what you do, and anything else you think we should know' {question}.

The second development was the tentative exploration of a new kind of transmedial hybridity, in which a single work was seen to exist *across* the digital/non-digital divide. There were two forces at play here, from opposite ends of our digital/non-digital divide: the creative interplay of non-digital forms and practices within digital works, and the inscription of digital forms and tropes within printed works.

Not everyone was quite so convinced by this transformation. In *Writing Spaces*, published in 2001, Bolter noted how print was becoming 'hypermediated', incorporating some of the aesthetics of the digital domain, 'in self-conscious imitation of and rivalry with electronic media, especially the World Wide Web' (2001, 46). The nouns 'imitation' and 'rivalry' indicate that the binary model was still very much to the fore in Bolter's thinking: 'hypermediation' was simply a pragmatic attempt by print media to keep hold of a readership in an increasingly digitised existence.

Yet, the hard truth was that Bolter's view was already out of date when it was written. Works such as *The Unknown*, *The Doll Games* and *House of*

Leaves show that, by the end of the 1990s, digital and non-digital authors alike were already starting to explore the symbiotic relationship between the digital and the non-digital in new and artistically dynamic ways. In the decade after Michael Joyce's *afternoon: a story*, the world saw the rapid growth of the laptop computer and PC, the world wide web, domestic internet connection, compact discs, the mobile phone, digital cameras, satellite television and wifi. The social and cultural impact of all these technological developments on ordinary women and men, over such a short period of time, was significant. In essence, lived experience was increasingly mediated between the digital and the non-digital. In other words, even by the late 1990s, the binary model was becoming irrelevant to what was actually happening in people's homes and in the workplace.

Taken from this perspective, works such as *The Unknown*, *The Doll Games* and *House of Leaves* can be seen as early explorations into this new digital/non-digital condition. As I've shown, both *The Unknown* and *The Doll Games* have many characteristics of classic postmodern hypertext fiction. Yet, they also exhibit elements that are less easy to situate within this tradition: *The Unknown* was written collaboratively by a small group of writers and involved both printed text and physical performance; *The Doll Games* elicited readers' experiences through an open invitation tacked awkwardly onto the story itself. There's a sense of frustration with both stories, a frustration with the limitations of technology at that time, but also a frustration with what hypertext fiction had become.

House of Leaves captures a similar sense of exasperation with the printed novel. Its use of hypertextual structuring and aesthetic tropes was not a simple case of rivalry or imitation of a more dominant form, as Bolter suggests. Instead, it was part of the same creative imperative that was apparent in the work of artists active in the digital domain: the desire to represent a new kind of hybridised experience. Yet behind all this lay something else too, far-more profound. As we've seen, works such as *The Unknown, The Doll Games* and *House of Leaves* had inscribed within them the death rites of postmodernism itself.

Hypertext fiction: death and rebirth

Hypertext fictions did not stop in the early noughties of course. Works such as *The Unknown, The Doll Games* and *The Disappearance of Miriam* mark a juncture, not an end point. Hypertext as a creative medium continues to be explored by writers of both print-based and digital works as an effective platform for storytelling. Yet, as we'll see, the how and the why behind these creations have noticeably shifted since the turn of the century. What we find is a new generation of writers for whom hypertext fiction is no longer about embracing a transformative, iconoclastic technology which threatens to usher in the end of print; instead it is much more about exploring the condition of a postdigital world.

As we've already seen with *These Pages Fall Like Ash* (2013) by Tom Abba and Duncan Speakman, *Seed* (2017) by Joanna Walsh and *Breathe* (2018) by Kate Pullinger, hypertext fictions quickly mutated in different ways, as writers both embraced new technological affordances but also began to move away from some of the creative assumptions that had underpinned earlier works. The exponential growth of smartphone ownership, coupled with the spread of wifi connectivity and increasing bandwidth, suddenly meant that writers could think of hypertext fictions, not as something viewed at home, on a clunky PC or laptop, but instead as a story accessed anywhere in the world. Added to this the opportunities offered by smartphone functionality such as GPS, the inbuilt camera, speaker and microphone, and the world of the simple 'link and node' hypertext fiction was transformed.

These are important developments and I'll be discussing them in detail in the next section of this book. What I'd like to do in this chapter is focus instead on how the form and structure of hypertext fictions actually changed in this later period and then relate this to the paradigmatic shift in creative modality that I've already identified. Stories such as *These Pages Fall Like Ash* and *Breathe* are not different from *afternoon: a story* or *The Doll Games* simply because of the improved level of technological sophistication available to their authors; they are also dissimilar because their authors have a different understanding of digital storytelling *per se*. By this I mean that authors such as Tom Abba and Kate Pullinger no longer see storytelling as being constrained within the ontological boundaries of postmodernism. Instead, digital authors increasingly see their work as exploring a new kind of creative modality, a modality that I have identified with the schematic of map/rhizome/string-figure. Central to this modality is a new kind of relationship with digital technology.

With the arrival of mobile technology such as the smartphone and tablet, the last vestiges of, what I have termed, the binary model of digital/non-digital domains, were swept away. In its place came a new hybridisation in which the relationship between the digital and the non-digital became symbiotic in the ways I have already started to discuss. This affected both digital storytelling and print-based media, just as we've already seen with *House of Leaves*. Both hypertext and print-based fiction therefore became expressions of and responses to a new socio-cultural condition, one which I have termed postdigital. Although the roots of this extend as far back as the late 1990s, it was only in the first decade of the twenty-first century that this new creative modality really began to come into its own.

A growing frustration with hypertext fiction was already apparent by the late 1990s. In an article written in March 1998 and entitled 'www.claptrap.com', Laura Miller noted, 'What's most remarkable about hyperfiction is that no one really wants to read it, not even out of idle curiosity' (1998, para. 3). Miller was writing back to Robert Coover's 1992 essay, 'The End of Books', which, as we've seen, had predicted the overthrow of the printed novel. For Miller, the idea that readers 'ought to be, and long to be, liberated from …

linear narrative and the author' (1998, para. 5) was an illusion. Instead, navigating the nodal structure of hypertexts was 'profoundly meaningless and dull' (1998, para. 8).

> If any decision is as good as any other, why bother? Hypertext is sometimes said to mimic real life, with its myriad opportunities and surprising outcomes, but I already have a life, thank you very much, and it is hard enough putting that in order without the chore of organizing someone else's novel.
>
> (Miller 1998, para. 8)

Although Miller was being deliberately provocative, her article did touch on some important truths: despite huge expectation, writing and reading hypertext fiction by the turn of the century remained a niche activity, stuck in what Pope calls 'a sort of twilight zone, apparently known only to a few insiders' (2006, 448). Kirby is even more severe, noting how 'the citizens of the hypertext community are, on the whole, an aging, nostalgic group' (2009, 222).

There are at least two aspects to this. The first was that a growing number of writers and theorists were recognising some of the issues raised by Miller. Rather than 'empowering' and 'liberating' the reader from any imposed meaning or structure, it was increasingly recognised that readers of hypertext stories tended to feel, what Ensslin identifies as, 'restricted, or disempowered, in their decision-making processes, a situation which more often than not results in feelings of frustration' (*Canonizing Hypertext* 2007, 3). Landow's concept of 'wreader' (1992) was becoming, at best, an idealised fantasy, and at worst, a misconceived understanding of what readers actually wanted.

In fact, as Ensslin and Skains, amongst others, recognise, the hypertextual interface, with its carefully planned architecture and navigational options, is as much about the overt manipulation and control of the reader (through the 'guard' function in Storyspace, for example), as it is about engendering wreaderly empowerment. If, as Bolter and Joyce claimed in 1987, '[a]ll electronic literature takes the form of a game, a contest between author and reader' (1987, 49), then it is a game in which the cards are very much stacked in favour of the author and where the reader can quickly become an unwilling participant.

In her book chapter 'The Interactive Onion' (2011), Marie-Laure Ryan outlines four levels of user interactivity that authors can adopt in their digital stories. They range from peripheral interactivity (level one) to real-time story generation (level four). I would argue that hypertext fictions are predominantly to be found at level two where, according to Ryan, interactivity is limited to the effect on narrative discourse and the presentation of the story (2011, 40). Here, we're very much in the territory of Joyce's exploratory hypertexts (1995, 41).

Ryan goes on to identify three narrative structures that support this level of interactivity – the sea anemone, network and tree – each one having a fundamental effect on the interplay between text and reader. In the sea anemone structure, all subnodes are connected to a central node. This modular design allows readers to probe deeper and deeper into the narrative, while always having the option to click straight back to the home page. A good example of this kind of hypertext fiction would be Robert Arellano's *Sunshine '69* (1996).[4] The sea anemone presents readers with a non-hierarchical choice of top-level nodes, from which radiates a series of hierarchical pathways. As the reader drills down along their selected pathway, they engage with subplots and seemingly inconsequential detail that slowly builds into a reassessment of the wider story.

In comparison, the network is the classic structure for hypertext fiction that we've already seen with works such as *afternoon: a story* by Michael Joyce or Stuart Moulthrop's *Victory Garden*. In its purest incarnation, there is no overall narrative pathway but instead a non-hierarchical series of nodes, offering a potentially endless series of unique pass throughs. In reality, however, authors of hypertext fiction adopt a more nuanced approach, modulating the complexity of network architecture, constraining and freeing the reader as deemed necessary across the narrative. Ryan uses the term 'loop' to describe the effect that a network structure brings to the reading experience.

> The formal characteristic of the network structure is the existence of loops that offer several different ways to get to the same node. These loops make it possible to circle forever within the network.
>
> (2011, 41)

These multiple pathways mean that the author can ensure, should they wish to, that it is not possible to predetermine the order in which certain lexia will be experienced by the reader, thereby having a profound effect on the reader's interpretation of the story. It also means that, in any pass through, a different story is created. Ciccoricco notes that these loops make repetition and variation key characteristics of these kind of digital stories, stressing the point that the hypertextual node is semantically different from the textual fragment (2012, 480). This, of course, is Lander's wreader at maximum effect, conjuring up their own unique story each time they navigate the network architecture of a hypertext fiction. It's also the structure behind all those issues raised by critics such as Miller and Kirby.

A more straightforward structure would be the tree. Here, a series of separate pathways (or branches) radiate out from a central node (or trunk). Once a branch is chosen, the reader moves along sequentially and is denied the option of leaping across to another branch. Without the network architecture, there are no loops – once a branch has been selected, there is no going back. Ryan compares this sort of architecture to the *Choose Your Own Adventure* children's stories where the branches correspond to a set of predetermined narratives from which the reader can only ever select one at any given time. As the reader moves along their chosen pathway, they are asked to make additional choices, leading to further branching. The benefits of this approach are that it offers the author maximum control over the narrative. Narrative coherence, structure and meaning are predetermined by the author, rather than being generated, at least in part, by the reader as they progress through network loops. However, the reader becomes far-more passive, merely clicking through a sequence of pages, one by one. The tree structure may offer a way of getting around some of the issues identified with classic hypertext fiction yet, as Hayles notes, 'it is also the least interesting, because the reader (or user) does not get more out of the system than what the author put into it' (2008, 44).

The sea anemone, network and tree architectures provide a useful way of conceptualising hypertextual interactivity (level two) by the beginning of the new century. Although each offers powerful ways in which the reader engages with the story, they also highlight something that was perhaps missed by the prognostications of early adopters such as Joyce and Bolter: that hypertext fiction ultimately relies on the planned and strategic manipulation of the reader.

This leads naturally into the second area of dissatisfaction, namely, the claims that hypertext fiction represented a radical break with print media. Indeed, for some commentators, the insistence of difference remains inherently flawed, reflecting the need to maintain the professional standing of electronic literature and its authors rather than any creative actualité (Aquilina 2018, 206). Indeed, as Hammond notes (2016, 159–160), some theorists were not slow in arguing that printed fiction could also be both non-linear and interactive in nature. Non-linearity, for example, has a long tradition within the western canon, where stories begin *in media res* and then proceed non-sequentially. Works such as B. S. Johnson's *The Unfortunates* (1969) take this to an extreme in terms of the physical ordering of the printed sections by the reader; yet, as Hammond notes, 'all printed books give their readers the freedom to skip from one section to another, to skim, to flip, to read out of order' (2016, 159). And, if readers are interactive with digital texts, then is the inference that they are therefore non-interactive with printed texts? As many critics have pointed out, reading any text, whether digital or non-digital, is an active process in which the reader co-creates any meaning. Currie uses the term 'pragmatic inference' to refer to this process, whereby the reader takes meaning directly from the words on the page but also infers lots of other stuff that isn't explicitly stated by the author (2010, 15). In other words, the reader is always interacting with the text, and the story is always a product of this two-way, 'pragmatic' interaction. From this perspective, Lander's notion of wreader is rendered tautological in the sense that readers are always involved in the active construction of a text.

A growing awareness of these limitations began to have a significant influence on hypertext fiction from the beginning of the new century. Writing in 2006, Pope recognised that a more balanced approach towards the design of hypertext fiction was needed. For Pope, as for a growing number of digital authors, the challenge was less about creating an avant-garde artform predicated on the uniqueness of the digital experience, and more about responding to the traditional expectations of the reader that were expressed by Miller. Pope recommends six key principles that, in essence, reintegrate the digital and printed novel (2006, 460–462). The first principle sets out his agenda: that readers' expectations should not be gratuitously frustrated. Quite simply, the effort the reader needs to invest in making sense of the story needs to be worth it. The moment the reader feels that this balance between effort and reward is skewed too much towards the author is the moment the reader gives up. The second principle is that the interface needs to make sense to the reader, particularly in terms of how to proceed and navigate. This doesn't mean to say that authors need to overwhelm the reader with navigational options. It might be that a minimalist approach is indeed far-more suitable to the intended story. However, it does mean that authors need to recognise that, unlike a printed book, a digital story can come with its own unique structure and aesthetics. This means the task of a reader of a digital text is doubled: not only to do they have to read and comprehend the written text, but they also have to work out the architectural and aesthetic rules of the story (and the possible relationship between the two). This can be exciting for the reader, of course, but it can also lead to the sort of cognitive overload that Ensslin and Skains identify (2018, 298). Once again, it's all about the trade off between effort and reward. One possible solution is to understand the interface as part of the narrative scaffolding that comes at the beginning of the story: just as context and backstory fades away as the reader progresses, then so too might the complexity of the interface. Pope's third principle is the avoidance of reader disorientation, both in terms of design (Where am I? What do I do next?) and plot (What is happening? What is the story about?). The balance between wreaderly empowerment and readerly frustration needs to be considered at all times. What is so noticeable about the hypertext gurus of the 1980s and 90s is their general reticence when it comes to talking about plot and story. And yet, just as the author of a printed novel supports the reader in their journey, even in the most complex of stories, then so too must the digital author. A hypertext fiction whose only raison d'etre is the exploration of postmodern complexity and indeterminacy is going to quickly alienate any reader because of that 'doubled' cognitive load. Fundamentally, the success of any hypertext fiction is based on the strength of the reader's conception of 'story'. As Miller says, 'story is fiction's trump card' (1998, para. 5). And in this, hypertext fiction is no different from any short story or novel ever written.

Fourthly, and building on the previous principle, the digital author needs to ensure both the flow and immersivity of the story. Interactivity can certainly support these qualities, but, as Pope makes clear, they can also needlessly disrupt the reader's experience when not aligned with the story. Fifthly, hyperlinks should be designed so that they always progress the story in a meaningful way that will engage the reader and further the story. Too many links that are essentially cosmetic, little more than dead ends, can kill off a story. As Daniel

Punday notes, meaningful progression is a critical part of a digital story's 'textual architecture' (2018, 144–145). Finally, Pope recommends that hypertext fictions, just like printed stories, offer some degree of closure. This doesn't have to be the classic denouement; it may well be that a 'traditional' end is not appropriate for the story; as Joyce noted in a discussion about his own hypertext stories, '[c]losure is, as in any fiction, a suspect quality, although [in hypertext fiction] it is made manifest' (*Of Two Minds* 1995, 186). However, there's a danger that hypertext fictions suffer from a kind of existential collapse at the very end, or leave their readers trapped in an infinite loop. Digital authors need to ensure effective closure no matter which pathway the reader has taken through the story.

So, with all of these caveats and recommendations in mind, let's look at a contemporary hypertext fiction and see how they compare to what we have already examined in this chapter. 'The Role of Music in Your Life' was written and published in 2016 by the online literary magazine, *Five Dials*, in partnership with the digital technology company, *Present Plus*.[5] Perhaps what is most striking about the story is its knowing use of hypertext as a cultural norm. 'The Role of Music in Your Life' mimics the sort of online survey or questionnaire that many of us have taken as we've been browsing the internet. The first page, #1 What is your favorite genre?, contains a single multiple choice question. All the reader has to do is select one of the six options. However, in reality, all these options lead to the next sequential page, #2 How do you listen to music?. This pattern repeats itself for the first five pages:

#1 What is your favorite genre?
#2 How do you listen to music?
#3 What kind of headphones do you use?
#4 What was your first instrument?
#5 What was your first album?

The reader, of course, is not aware that the options they make have no impact on the next question. Instead, they assume that the questionnaire is responding to their selections, that the software is somehow listening to them, modifying the next question based on their previous answers. Importantly, the authors have ensured that readers cannot go back to a previous page to test the mechanics of the story; once an option is selected it's as though a door is closed behind them. The only options are to keep going forward or to come out of the story completely.

From hereon in, things start to change. Page {#6} asks: 'You're hungry. You're going to eat…', the first hint that this questionnaire on music is not all that it seems. After page {#8}, 'How did you mourn the death of David Bowie?', one of Bowie's iconic lightning bolt graphics flashes momentarily on the screen. And then, suddenly, on page {#13}, things shift again.

It's now clear that the reader is not in fact completing an online questionnaire but is actually sitting in a room in a house, speaking to a mother about giving piano lessons to her son, a sixteen-year-old boy named Carl. The structure of the questionnaire falls away and instead the options represent a conversation. In fact, it's clear that the questionnaire was somehow never really a questionnaire as the reader thought, but was bound up in this strange scene that is only now being revealed. The rest of the story plays out rather like a piece of contemporary flash fiction. It is all dialogue-led, apart from occasional graphics and video, which burst onto the screen for less than a second, momentarily puncturing the physical textuality of the storyworld. The fact that the reader is addressed directly gives the story a feel of the sort of second-person dramas popularised by short story writers such as Lorrie Moore (*Self Help*, 1985), where the reader becomes a mixture of both themselves and the character they are playing in the story (in this case, a piano teacher). There's also more than hints of postmodern short stories such as J. G. Ballard's 'Answers to a Questionnaire' (1985) and Lydia Davis' 'Jury Duty' (2001) in which the story is structured around the answers to an omitted series of questions.

'The Role of Music in Your Life' conforms to Pope's principles outlined above. The reader is neither exasperated nor disorientated by the design or interface. Navigation is intuitive and, indeed, part of the story. Links always progress the story and there is an effective ending. It should be stressed that these conditions do not militate against complexity of experience for the reader. As I've said, 'The Role of Music in Your Life' can be understood as a piece of interactive flash fiction. In that sense, the story is enigmatic and self-knowing, where the real story only emerges through subtext. Is Carl actually being abused? Are his mother and father in a loveless marriage? Is the mother flirting with the reader? And what happened to the previous piano teacher? None of these questions are directly addressed, but rather hover, unstated, on the edges of the drama, unsettling the strained calm of the dialogue: 'I worry [Carl's] reaction to music is BEYOND PASSION. Though I don't want to scare you off' {#47}.

What's also interesting about the story is its structure. 'The Role of Music in Your Life' consists of 89 separate pages or lexia, including four images and two videos. However, despite this complexity, the structure is completely linear; at no point do any of the options have any effect on the story other than by sequentially progressing the reader to the next page. All the optional links on every page could simply be replaced by a single 'next' button. It's Hayles' tree structure but with no branches, just one big trunk. Or rather, its Hayles' tree masquerading as a network. In that sense, 'The Role of Music in Your Life' could be seen as a con or sleight of hand. But actually, from the reader's perspective, it doesn't feel like that at all. The interactive construction of a story is provided by the reader's expectation of that affordance. By inhibiting the return to previous lexia, the authors have deliberately stopped the reader from discovering that this expectation is not what is actually happening: that, in reality, they are simply clicking sequentially through 89 lexia.

What we see in 'The Role of Music in Your Life' is a particular form of pragmatic inference. Although Currie didn't discuss digital storytelling, it is clear that the reader's inferences regarding the underlying functionality of the interface are just as important as their inferences regarding the story itself; in fact, in work such as 'The Role of Music in Your Life', interface functionality and story are one and the same. The story is a compelling hybrid of the digital and non-digital, merging together the online questionnaire with the narrative conventions of print-based flash fiction. 'The Role of Music in Your Life' therefore follows in the footsteps of work that explicitly seeks to subvert the forms and structures of an increasingly normalised digital landscape. Such material forms a wider subculture of work written by writers seeking to satirise and parody the pervasive nature of digitality. This includes the writing of fictionalised reviews on the Amazon website (Phillips and Milner 2017, 27) to the parodying of confessional blogs, tweets and Facebook entries that get embedded in the blogosphere.

This wider role of digital storytelling within an activist or political agenda forms an important part of where hypertextual storifying is today. In the case of the research agency, Forensic Architecture, for example, the construction of hypertext narratives remains a key part of their output (Weizman 2017).[6] Forensic Architecture defines not only the name of the agency but also its innovative methodology. In the words of its founder, Eyal Weizman, Forensic Architecture 'refers to the production of architectural evidence and its presentation in juridical and political forms ... we seek, in fact, to reverse the forensic gaze and to investigate the same state agencies – such as the police or the military – that usually monopolise it' (9). Central to this approach is the placing of localised incidents within their wider social and political context, 'reconnecting them with the world of which they are a part' (9). The agency is highly interdisciplinary, consisting of architects, artists, technologists, journalists and filmmakers. It undertakes investigations at the behest of international prosecutors, human rights organisations and political and environmental justice groups, including Amnesty International and Human Rights Watch.

Forensic Architecture has two distinct imperatives. The first is the rigorous collection and analysis of data relating to a specific event; and the second is the presentation of that data in new and dynamic ways to the wider world, including both digital (web-based storytelling, for example) and non-digital (exhibitions, talks and performances). What makes Forensic Architecture (FA) particularly unique is its critical and creative engagement with the contemporary media landscape. As the agency say on their website:

> The premise of FA is that analysing violations of human rights and international humanitarian law (IHL) in urban, media-rich environments requires modelling dynamic events as they unfold in space and time, creating navigable 3D models of sites of conflict and the creation of animations and interactive cartographies on the urban or architectural scale.
>
> (Forensic Architecture)

The context within which Forensic Architecture operates is underpinned by the ubiquity of digital data, from CCTV, satellite imagery and real-time data flows, through to individual image and video capture. These are supplemented by non-digital forms, such as personal and community testimony. Such data are brought together so that particular events can then be remodelled spatially, creating web-based interactive representations of those occurrences over time. The ultimate aim of Forensic Architecture is to seek accountability for the violations and atrocities they investigate. As a consequence they place high priority on both the efficacy and reach of their work.

In 2016 Forensic Architecture was commissioned by Amnesty International to assist in the reconstruction of the secret Syrian detention centre at Saydnaya. The prison has become notorious for its incarceration of opponents of the Assad regime, a situation which has only worsened since the beginning of the Syrian crisis in 2011 (Weizman 2017, 85). The Syrian Government operates Saydnaya in secrecy: there are no visits from independent agencies and no reports or photographs. All that persists beyond the prison walls are the memories of those victims who passed through it without losing their lives.

In 2016 researchers from Amnesty International and Forensic Architecture used the testimony from survivors to help reconstruct the detention complex and the lives of those confined within it. *Saydnaya: Inside a Syrian Torture Prison* (2016) is a web-based, interactive story that allows users to experience the realities of life inside the prison.[7] For the detainees, their assistance in the reconstruction of the complex was very often the first time they had properly tried to understand what had happened to them. As Weizman explains, the co-construction of the virtual story was itself a form of therapy for the five participants, Samer, Dia, Jamal, Salam and Anas. *Saydnaya* includes video in which participants are seen working with Forensic Architecture in an iterative process, as preliminary modelling surfaces more detailed memories. In this sense, the methodology of the reconstructive process is as much a part of the story as the final virtual experience. The story is built around both individual testimony and the architecture of the detention centre (shown in Figure 5.4) and the user has the option of exploring the story by either form (Explore by Location; Explore by Witness). Much of the prison was unlit and underground; movement through the prison was done while the prisoners were blindfolded (88). Any speaking was prohibited. One of the witnesses describes how once, while being led from his cell, he was forced to hold his hands over his eyes. As he was being hit by a guard, his hand momentarily fell away and he caught sight of a huge circular space filled with cells. 'The instantaneous sight he gained was a momentary leak of vision into a spatial perception otherwise fully defined by sound' (Weizman 2017, 93). The aesthetics of *Saydnaya* are therefore dark; sounds are infrequent and belie both the architecture of the prison and the particularities of the repression within it. The story progresses through spatial exploration, video, text and sound.

Returning to Hayles' taxonomy of structure, I would suggest that *Saydnaya* has level two interactivity and a sea anemone architecture by which the reader

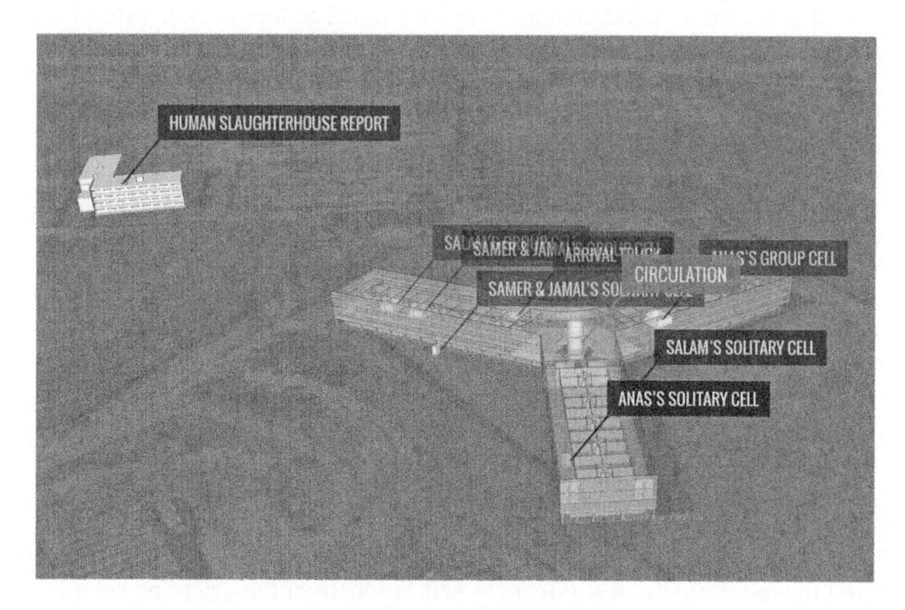

Figure 5.4 Saydnaya: Inside a Syrian Torture Prison (2016). Screenshot.
Source: © Forensic Architecture and Amnesty International 2016.

can always return to the main menu from wherever they are in the story. It does not show physical violence or assault; instead, it slowly immerses the user in the prison's psychology of fear: the ever present threat of violence and death, and the slow passing of time in the darkness of the cells.

Clearly, the intention of *Saydnaya* is not to entertain. Instead, the story seeks to raise awareness of human rights abuses and to put international pressure on the Syrian government to allow independent monitors into its detention centres. One of the options in the horizontal menu is 'Take Action'. From here the user is invited to send an email to the International Syria Support Group, pledging their support for independent monitors.

Saydnaya was 2016 winner of the *Peabody-Facebook Futures Media Award* for best interactive story and was one of the 2016 winners of the *Digital Dozen: Breakthroughs in Storytelling*, awarded by the Columbia University Digital Storytelling Lab. Yet, if *Saydnaya* is a hypertext story, then it is one in which the hypertextuality of its form plays a purely functional role. It is the story that matters. As a web-based experience, *Saydnaya* can be accessed freely across the globe. It has also been part of numerous exhibitions and conferences, including *Beautiful New Worlds: Virtual Realities in Contemporary Art* exhibition at the Zeppelin Museum, Friedrichshafen, Germany. While it is interactive, a balance is maintained between the freedom of the user to explore, and the narrative requirements of the authors. Unlike *afternoon: a story*, *Saydnaya* has a story it wants to get across and again, unlike *afternoon: a story*, the story does not involve any overt response to hypertextuality as a metaphor for a postmodern ontology. Instead, *Saydnaya* wants to prioritise affect through the recreation of a situated and embodied encounter with the real world. And as such, Pope's six key principles are all in evidence. As Weizman states, human rights groups have a long history of using 'the affective power of the arts in helping stir public compassion' (2017, 94). Yet, for him, Forensic Architecture has intensified this relationship, exploring new forms of artistic practice to present both the 'modes and the means by which reality is sensed and presented publicly' (94). *Saydnaya* does not seek to problematise notions of reality, what Danius called the representation of 'authentic experience' (2002, 3). Instead, its priorities lay elsewhere – with affect and embodied meaning, told through the openly-accessible, transmedial affordances of hypertext storytelling.

While works such as *Saydnaya* are professionally produced and exhibited, there are other contemporary trends that have ensured hypertext fiction has also witnessed a resurgence in the home and classroom. New platforms such as Twine (2009) and Genarrator (2009) have opened up the creation and publication of hypertext fiction to anyone with a computer and access to the internet. Ensslin and Skains are surely right to highlight that this development is just part of a wider movement within hypertext storytelling towards collaborative and interdisciplinary practices more generally (2018, 296).

As we've described, the world of hypertext fiction throughout the 1990s and into the new millennium was very much restricted to an elite of avant-garde artists and practitioners, some of whom we've already looked at in this chapter. Hypertext fiction offered a new kind of creative opportunity that would soon overthrow the domination of the printed book. Any artistic interdependence between the non-digital and digital domains was downplayed, safeguarding electronic literature as an institutionalised field of study and practice (Aquilina 2018, 206). Storyspace (1987) was a critical part of this. Developed by Jay David Bolter and Michael Joyce, Storyspace dominated the world of hypertext fiction and continues to remain a significant platform today. However, Storyspace is not free; a single license is currently over £100. This cost, and the overall complexity of the platform, ensured that Storyspace remained a specialist tool, one dominated by first-generation hypertext authors such as Shelley Jackson, Michael Joyce and Robert Coover. Things only started to change with the release of new platforms such as Twine and Genarrator as well as social media technology.[8] The exponential growth of internet-enabled computers in the home, opened up a new and expanding market for web-based, hypertext fiction. Both these platforms are free; Genarrator, developed by James Pope and James Ready at Bournemouth University, also hosts the completed stories, making them publicly accessible should the author wish it (Pope 2013). Whereas Genarrator supports a variety of multimedia, including video and sound, Twine is very much text-based. As open source, Twine remains the product of its community of users. Authors can either use it as a web interface or as standalone software downloaded to their computer.

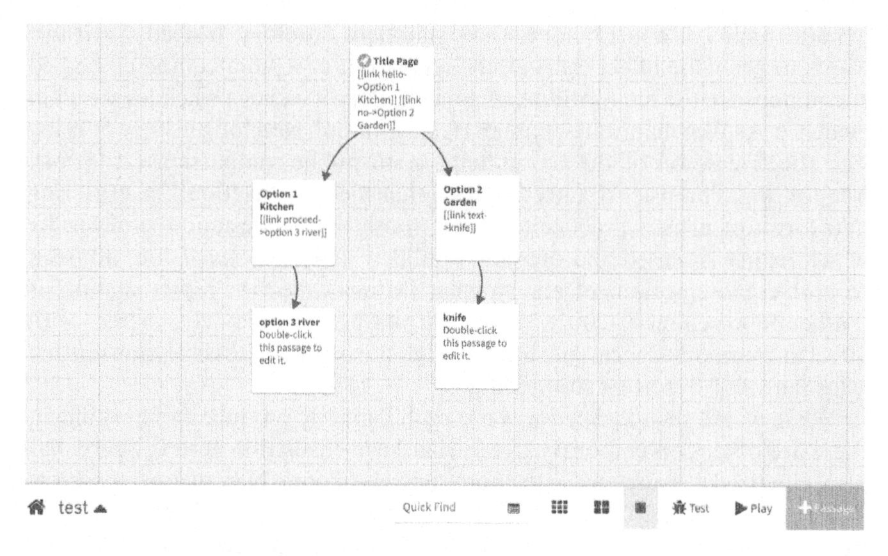

Figure 5.5 Twine interface. Screenshot.
Source: © Chris Klimas.

By democratising the authorship of hypertext fiction, platforms such as Genarrator and Twine have overseen the growth of, what Ensslin and Skains describe as, 'highly personalized and personalizable, autobiographical forms of hypertextual writing' (2018, 295). Many of the stories, for example, show a disregard to the postmodern imperatives of early hypertext fictions. Instead, these new platforms support a more straightforward approach to narrative structure, foregrounding creative practice as an affective act. The stories that emerge are less about the exploration of technology as a distinctive domain, and much more about understanding the interplay between the non-digital and the digital, the human and the technological. Despite the web being almost ubiquitous in our lives, our ability to creatively engage with hypertext is still very much constrained. Although the millennial generation are often extremely confident when it comes to social media, for example, their understanding of how a web page actually works, its underlying code and functionality (that four layer model I presented in Chapter 3, Figure 3.5), can be limited. Platforms such as Twine and Genarrator provide a safe environment in which this facade of hypertextuality can be deconstructed. Work can be co-developed across a team of contributors and then published to the web and reviewed by others. The emphasis is less on narrative sophistication *per se* (although this can also be present, of course) and more on the personal or collective exploration of digital technology as a means of telling stories. In other words, this is exactly the sort of autofictional writing that Gibbons identifies as being a key metamodernist trope, exploring 'the sociological and phenomenological dimensions of personal life … and how experience is often mediated by textual and/or digital communication' (2017, 130). Very often these are stories that draw overtly on the traditions of print media in both form and narrative design, providing a radical new space of postdigital experimentation in which the affective embodiment of experience are paramount. As Ensslin and Skain note in their discussion of Twine, such platforms reach out to areas of the population previously excluded from hypertext writing, pushing it 'outside of the experimental art realm and into the mainstream' (2018, 303).

What I've examined in this section has been the slow transformation of hypertext stories since the turn of the new millennium. Although I've explicitly examined two stories here – 'The Role of Music in Your Life' (2016) and *Saydnaya: Inside a Syrian Torture Prison* (2016) – much of what I've said also relates to the stories that I've discussed in previous chapters from this period, including *These Pages Fall Like Ash* (2013), *Seed* (2017) and *Breathe* (2018). Critical to my argument has been the recognition that these postmillennial stories represent a turn away from postmodern angst towards new forms of embodied connectivity. Or, as I've been describing it throughout these chapters, from the labyrinth to the affectivity of Haraway's string figures. If hypertextual works such as *Saydnaya* offer new forms and modes of 'troublemaking', then platforms such as Twine and Genarrator open up digital storytelling to anyone with a computer. While Pope's six principles (2006) demystify what narrative affectivity might actually mean for the hypertextual story, Twine and Genarrator are equally important in the way they offer (co)-creative and critical engagement with digitality *per se*. As I made clear in Chapter 3, it is this opportunity to *become other-wise* (Rob Pope 2005, 29), this *worlding of the world*, that is such a vital function of our map/rhizome/string-figure modality.

Hypertext and the novel today

If hypertextuality has gone through a transformation in the digital domain, then it has also continued to have a significant effect on the non-digital too, particularly the printed novel. We've already seen how works such as *Gravity's Rainbow* (1973) and *House of Leaves* (2000) captured wider changes to our creative modality, from the postmodernist denial of self (Derrida 1978) to a more nuanced position, in which, to use Mary K. Holland's phrase, 'much-missed portions of humanism' such as 'affect, meaning and investment in the real world' (2013, 8) are salvaged from the postmodern desert.

I would argue that this phenomenon has only deepened. One of its more telling features is the increasing entanglement of the digital and non-digital domains. While Danielewski was one of the first authors to explore the creative interplay between the web and the physical novel, this has now become mainstream. There are two broad approaches to this. The first, and the more common, are the authors for whom the digital domain supports the creation of paratext (as opposed to a simple author website) which directly complement, enhance and embellish the primary work. One of the best examples of this is J. K. Rowling's *Pottermore* website.[9] Released in 2011, *Pottermore* supports a rich digital enhancement to the novels. As well as providing a commercial platform, *Pottermore* also contains additional material on various aspects of the Harry Potter world, including spells, characters and creatures. The first incarnation of the website (up to its relaunch in 2015) was highly interactive, inviting readers to set up a *Pottermore* account from which they could then create spells and potions and compete in the House cup.

Richard House's *The Kills* (2013), on the other hand, takes a different approach: here the paratext consist of forty specially-produced videos. In the ebook version of the novel, these are accessed through a hypertext link to a hosting website.[10] In the print version, the unique URLs for each video are listed at the very end of the book, along with the following advice:

> Richard House has also created multimedia content that takes you beyond the boundaries of the book and into the characters' lives outside its pages ...This additional content is freestanding and can be experienced before or after reading the book.
>
> (2013, 1001)

I would argue that both J. K. Rowling and Richard House are examples of authors exploring the creative potential of the postdigital condition. The primacy of the printed novel for these authors remains intact; instead, they have sought a different kind of mediation with the digital, a network of digital and non-digital paratextical nodes, with the printed novel at its centre. In Rowling's case, the original version of *Pottermore* offered a form of immersivity for her readers, an opportunity to step into the world of Harry Potter and interact with key elements of the stories. Similar to the *House of Leaves*' website, Pottermore also offered discussion boards as a way of encouraging a sense of community across her readership. In contrast, House used digital content specifically as a means of supplementing his novel. The videos are short in duration, lasting between one and two minutes; they are all shot on location, often through handheld cameras, which gives them a strong sense of realism. The majority have overlaid text, which either represents a voice talking to us, or the transcript of emails or web chats. Some, however, have recorded voices (Watts, for example, who is a character in the novel, or Rem's sister, who is not). Like *The Kills* itself, the videos are enigmatic, offering a fragmentary and partial glimpse into the story, capturing the sights and sounds of the various locations, such as Iraq. Yet, in the very digitality of their form, and their self-conscious play with the forms and codes of mediatised communication more generally, these videos foreground the globalised, technologically-enmeshed lives of the novel's characters. It's this sense of authenticity that is the videos' most compelling characteristic for the reader, authenticity both in terms of design but also their overall effect on the story. If these digital paratexts have a role, it is one that firmly anchors the novel in the domain of a situated and embodied realism.

The second broad approach is where the forms and codes of the digital are brought directly *into* the printed text, in other words, not as paratexts. *theMystery.doc* by Matthew McIntosh (2017) is a good example of a contemporary novel that explicitly seeks to explore postdigital poetics through the printed page. At 1,653 pages and weighing over four pounds, it would be hard to think of a reading experience further removed from a smartphone or iPad. In its very physicality, *theMystery.doc* seems to defy the encroachment of digital technology on the publishing industry. And yet digital technology is pretty much everywhere in the novel, including the title. In the best traditions of metafiction, the main character is an author whose latest novel is being written in a word file called theMystery.doc. This embedded work is described as 'a post-post-neo-modern mystery story' (2017, 51) which also, of course, represents McIntosh's novel. Whereas *The Kills* maintains a traditional third-person narrative perspective throughout the printed novel, pushing any formal experimentation (first-person voice, email, webchat) out into the paratextual videos, *theMystery.doc* works in the opposite manner, deliberately drawing such experimentation *into* the printed text. The effect of this is to create a compelling kind of metafiction, drawn together from a vast range of digital *and* non-digital fragments including text, photographs, machine code, audio transcripts and film and video frames. The positioning of sequential frames on adjacent verso or recto reproduces a flick book effect, in a similar way to how Jonathan Safran Foer used images of a 'falling man' at the very end of his novel, *Extremely Loud and Incredibly Close* (2005). There is an ebook version of *theMystery.doc* but it remains faithful to the functionality of the print version, with no digital enhancement. From this artistic decision, it seems safe to assume that, for the author, the rendering of digitality within analogue form remains a critical part of the work.

In this sense, *theMystery.doc* has more in common with *House of Leaves* (2000) than *The Kills* (2013). Yet the mediality of *House of Leaves* (2000) is far-more circumscribed than *theMystery.doc*. As I've shown, *House of Leaves* consists of textual (printed) documents, some autobiographical (Johnny Truant), others textual responses to a lost film (Zampanò's *The Navidson Record*). I have also argued that the *House of Leaves* does not exist as an ebook for this very reason: the printed (physical) nature of these textual elements are a critical part of the story. None of these elements purports to be anything other than written, textual testimony. This is in contrast to *theMystery.doc* which intermingles printed-based representations of online, real-time chat, classic first-person narration, film, video and photography in a sprawling kaleidoscopic mélange. Whereas *House of Leaves* was written for the savvy hypertext surfer of the 1990s, *theMystery.doc* is a novel for the smartphone generation who cruise seamlessly between the digital and non-digital domains in, what Adriana de Souza e Silva has called, 'hybrid space' (2006).

However, *theMystery.doc* is also a product of postdigitality in other ways. I have already argued in Chapter 3 that, in our postdigital age, printed books have become signifers of functional deficit: in other words, they have come to represent 'that which is not digital' rather than any natural or intrinsic form of published work in and of itself. This makes works such as *theMystery.doc* particularly complex: the novel represents transmedial digitality in a form which ultimately decries its ability to do so. It would have been easy enough to bring some of this functionality into the ebook version of course, but as I have already shown, the author has refused to do this.

This insistence on the purity of the reader's analogue experience of the novel turns *theMystery.doc* into a kind of textual trompé-l'œil. Trompé-l'œil is French for 'deceives the eye' and defines an artistic approach that creates the illusion of a real, three-dimensional object or scene within a painting. Like classic trompé-l'œil paintings such as Edward Collier's *A Trompe l'Oeil of Newspapers, Letters and Writing Implements on a Wooden Board* (Figure 5.6), *theMystery.doc* can be considered an artistic illusion which draws attention to its own artificiality as it reaches for verisimilitude. Just as trompé-l'œil painters such as Edward Collier rendered three-dimensional effects within a two dimensional media form (the painter's canvas), so McIntosh is representing digital effects within the non-digital domain.

By pulling these digital forms into the novel, McIntosh defamiliarises them as texts, thereby emphasising their constructed nature. For example, video and film are reduced to inanimate stills that only 'come alive' when the reader flicks through the pages; real-time web chat becomes a kind of prose or concrete poetry, sprawling across the page. This materiality of the printed work remains a fundamental experience of the story: the reader's physical action of turning each page, of holding all 1,653 pages in their hands as they do so. The reader is never allowed to forget that it is their own bodily entanglement with the book that ultimately creates the story. If *House of Leaves* is a networked novel, a nodal constellation of independent paratextical documents, then *theMystery.doc* represents what Tim Ingold conceives as a 'meshwork' of interdependent relations, or, to use his own term again, 'interwoven lines' of entanglement, rather like Haraway's string figures (2016, 106). A recurring scene across the novel is an online discussion between Michele, a website greeter for WebsiteGreeters.com, and a series of customers – on each occasion, the customer is consumed by the doubt that they're not talking to a real person:

VISITOR 1969: Is this an automated program or is there a live person on the other end?

MICHELE: I am a live person

(2017, 30)

The artificiality of these digital inscriptions become metaphors for the artificiality of all media forms; yet what haunts *theMystery.doc* is not Baudrillard 'desert of the real' and its loss to simulacra (1983), but rather the human

Figure 5.6 Edward Collier, *A Trompe l'Oeil of Newspapers, Letters and Writing Implements on a Wooden Board* c.1699 (T03853). Oil paint on canvas.

Source: © Tate, London 2019.

striving for meaning and understanding. At its heart the novel is a search for authenticity in a world where existence is fractured across media domains. If *theMystery.doc* is a postdigital network, then its central node is the reader, and its focus the embodied and affective striving for truth.

Hypertextual storytelling/research

The early pioneers of hypertext stories were as much researchers into new forms of creative expression as they were avant-garde authors. As we've seen, a number of them, including Michael Joyce (1995) and Jay David Bolter (2001), published significant works about their own practice in which they openly discussed both their own approach to digital storytelling but also what they felt were its wider ramifications. From these works it is clear that they saw the new media of hypertext as offering the opportunity to explore the creative limits of poststructuralist theory: in other words, hypertext stories became a practice-based mode of research into the condition of postmodernity. Here, stories such as *afternoon: a story* and *Patchwork Girl* explored ways by which the affordances of hypertextuality could both foreground and in some way respond to, hegemonic power structures. As I've mentioned, there was a certain duality to these stories, the sense that they were not only seen by their authors to be irredeemably enmeshed in depthless relativity and meaninglessness, but also, somehow, offered an empowerment of the wreader and a way through the labyrinth to some kind of insight.

The pivot away from postmodernity to the creative modality of the map/rhizome/string-figure brought with it a readjustment in how hypertextual storytelling could be used as a form of research. As Keri Facer and Kate Pahl have noted, there has been a steady movement within academe towards impactful, affective research activities, in which Facer and Pahl identify interdisciplinarity and collaboration as key characteristics. Suddenly, concepts such *materiality and place*, *praxis*, *storifying* and *embodied learning* have become recognisable tropes within arts and humanities research projects (2017, 17). In this new ontological paradigm, hypertextuality has become less a metaphor for postmodern entrapment, and much more a symbolism of new kinds of affective and embodied connectivity. From this perspective, the hypertext story can be considered as a form of postdigital wayfinding, a string-figure of co-creation, offering news ways of becoming (Rob Pope 2005, 17), new opportunities of becoming otherwise. As Bruno Latour expounds in his book, *Down to Earth: Politics in the New Climatic Regime* (2018), the imperative of today is to work out ways by which we can maintain co-existence on a fragile planet: '[i]t is not a matter of learning how to repair cognitive deficiencies, but rather of how to live in the same world' (25). If this 'deficit in shared practice' (25), as Latour calls it, is at the heart of Facer and Pahl's taxonomy of attributes, then so too can it be found in the form, structure and purpose of hypertextual stories in arts and humanities research projects.

As we'll see when we turn to locative mobile storytelling (Chapter 6) and collaborative, community-based storifying (Chapter 7), postdigital storytelling is already playing a significant role in the exploration of these imperatives. The 'in vivo' (within the living) of transdisciplinarity remains a critical part of this, a philosophical approach that draws strongly on Martin Heidegger's concept of 'the worlding of the world' (Bolt 2011, 87). Here, works such as *These Pages Fall Like Ash*, *Seed* and *Breathe* are reconfigured as innovative forms of postdigital wayfinding in which embodiment and transmediality remain overt within the narrative process. Concurrently, works such as *Saydnaya*, and the hypertext platforms, Twine and Genarrator, have helped in positioning biographical, autobiographical and autofictional narratives as an important and evolving genre within hypertextual poetics. Here, the actual making of the stories, rendered explicit in the case of *Saydnaya* through its inclusion in the narrative itself, becomes a key aspect of the praxis. Ruth Tringham's *Dead Women Do Tell Tales* project, Lawrence Cassidy's *Cultural Memory Project* and Karen Smyth's *et al. Preserving Place: A Cultural Mapping Exercise* are excellent examples of how these approaches can become a critical part of academic research activity and output.

Summary or what we talk about when we talk about hypertext

It has become *de rigueur* in some quarters to talk about the irrelevance of hypertext fiction, pushed 'into the footnotes of literature by the onward speed of change' (Kirby 2009, 223). Yet, to paraphrase Mark Twain, the reports of its death are surely greatly exaggerated. As this chapter has shown, it is not so much hypertextual storytelling that has died, but rather the postmodern assumptions from which it originally grew. In the new creative modality of map/rhizome/string-figure, hypertext fiction is very much alive and kicking.

Both this and the previous chapter, have followed the development of hypertext fiction from its first emergence in the mid 1980s, through to where it is today. This was a period that saw significant technological change, of course, including the rise of the personal computer and the smartphone. Yet, as we saw during earlier periods of technological upheaval, the impact that such innovations have on our lives are complex and profound, often prefiguring deep and sustained cultural and social transformation. The 'crisis of the senses' that underpinned modernist literary experimentation in the first half of the twentieth century was in part a response to the disruptive effects of the telephone, radio and cinema.

The way artists continue to respond to ongoing social, cultural and technological transformation is at the heart of this chapter. As we've seen, the rise of hypertext fiction coincided with the high-water mark of postmodernism. The fractured, labyrinthine structures offered by hypertext were an ideal form by which the most fundamental tenets of postmodernism could be explicated. Poststructuralist notions of 'the death of the author' (Barthes 1967), for example, were a direct influence on Lander's concept of 'wreader', while hypertext authors such as Michael Joyce and Shelley Jackson openly embraced narrative non-linearity and indeterminacy. Such hypertextual hubris naturally led to Bolter's declamation of 'the late age of print'; yet such comments also drew on the digital/non-digital binarity that lay at the heart of postmodernism.

Of course, this is not how things turned out. By the beginning of the new century, not only was it clear that hypertext fictions such as *afternoon: a story* (1987) and *Patchwork Girl* (1995) had failed to have much of an impact on mainstream reading habits, but it was also apparent that postmodernism itself was in terminal decline. Even by the late 1990s, it was evident that the ontological irony and scepticism that characterised postmodern literature was being superseded by a new kind of affective earnestness: a belief in the need to connect and to change. As David Foster Wallace stated in an interview in 2010:

> There's a certain set of magical stuff that fiction can do for us … One of them has to do with the sense of … capturing what the world feels like to us, in the sort of way that I think that a reader can tell 'Another sensibility like mine exists. Something else feels this way to someone else'. So the reader feels less lonely …
>
> (Lipsky 2010, 39)

As we've seen, although *House of Leaves* (2000) had many characteristics of a classic postmodern novel such as *Gravity's Rainbow* (1973), it was not a straightforward postmodern novel in and of itself. *House of Leaves* was less about the denial of truth and certainty and much more about the search for, what Holland calls, a 'space for an alternative to a postmodern insistence on ironic disaffection and meaninglessness' (2013, 119). In other words, *House of Leaves* marks the turn towards metamodernist affect.

Yet it also does something else too. *House of Leaves* can be viewed as a node within a much larger paratextical network which included both digital and non-digital elements. Indeed, the aesthetics of the hypertextual network were brought directly onto the printed page itself. If *House of Leaves* is metamodernist in philosophical position, then it is also one that breaks with, what had been, the traditional model of digital/non-digital binarity, albeit at a time when the non-digital domain was still dominant. *House of Leaves* was yet further evidence that Bolter's late age of print was not going to plan, at least in the way those early digital authors predicted. Instead, as Ted Striphas notes, the world had entered a more dynamic and open-ended age, characterized by both permanence and change (2009, 175). As *House of Leaves* demonstrates, central to this creative disruption were new and evolving forms of connectedness and transmedial hybridity.

With the new century, hypertext fiction drifted further and further from the beliefs and principles that oversaw its birth. No longer could it be described as the oeuvre of an avant-garde elite; instead, through freely available platforms such as Twine and Genarrator, hypertext fiction became democratised, a form in which almost anyone with a networked computer could participate. In these open-source communities, hypertext fictions have become far-more participatory and collaborative in their construction (Ensslin and Skains 2018). Postmodern disorientation has been replaced by a kind of postdigital autofiction: in other words, the works are less about exploring narrative indeterminacy or 'tricksiness' and much more about the creation of an affective, situated experience for the reader. Such open platforms also foreground authorial creativity as an act of empowerment in and of itself, inviting us to explore our embodied (often passive) entanglement with a digital domain which, in its pervasive ubiquity, has become invisible. The research agency, Forensic Architecture, is a good example of how these values have attained international reach and impact. For its founding director, Eyal Weizman, hypertext storytelling offers a powerful way to campaign, inform and educate. Award winning works such as *Saydnaya: Inside a Syrian Torture Prison* bring together both non-digital data (human testimony and physical evidence) with digital data generated by our mediatised landscape. The effect is both personal and embodied, but also contextual, analytical and explanatory. As with the autofictions of Twine and Genarrator, *Saydnaya* is as much a story of how it was made (including the relationship between evidential testimony and trauma), as it is an objective representation of what actually goes on inside the prison. And we must also remember that the locative mobile storytelling discussed in Chapter 6, and the collaborative narrative forms of social media analysed in Chapter 7, are yet further contemporary variations of the hypertextual story.

All these approaches give credence to our revised definition of creativity: a situated and embodied act, in which, through the (co-)creation of new knowledge, perceptions and understanding of the world are changed. It is also clear that the new creative modality of map/rhizome/string-figure is very much in evidence in these and other hypertextual works, outputs in which a new kind of ethical force and truthfulness are apparent. As Hammond states:

> The most fascinating interactive fiction being written today does not do what Bolter or Landow predicted it would. Instead, it does what literature has always done: it finds new ways to bring readers more directly in contact with what it means to be alive in their place and time.
>
> (2016, 174)

It's worth reminding ourselves of Harriet Hawkins'three critical geographies of creativity discussed in Chapter 2: creativity as an embodied, material and social practice; as inherently political; and as 'as a force in the world' (2017, 12). The overlap with what I've been arguing in terms of hypertextual storytelling is clear. Yet, for contemporary authors, something else has happened too: just as hypertext fiction has moved away from being a gated community of elites, so hypertextuality itself is no longer a straightforward digital enclave. If hypertextuality is an important aspect of contemporary life, then it is one which is increasingly hybridised in nature, spanning both digital and non-digital elements in increasingly complex dependencies. *These Pages Fall Like Ash* (2013) is a classic example of how digital authors are exploring this new hybridic terrain, as is J. K. Rowling's *Pottermore* website and Richard House's use of video.

Such hybridity is at the heart of what I have termed postdigital. As I have already outlined in Chapter 3, one critical condition of the postdigital is not simply a growing enmeshment of the digital and non-digital domains; encoded within it is also a fundamental shift in the cultural position of the printed book, from dominant to subordinate. In the digital age, the printed book was socially and culturally hegemonic; it was the digital book, the hypertext fiction, that was the interloper. At this time, the digital work was always in deficit, always being defined by that which it was not – the printed book. And crucially, of course, this functional deficit underpinned both the author's and reader's experiences.

In the postdigital age, however, this condition has been reversed. Beneath the growing hybridity of creative work, the hegemonic condition of print is being overturned. It is now the printed book that is in functional deficit; it is now the printed novel which is defined by that which it is not – the online, interactive, participatory story. For works such as McIntosh's *theMystery.doc*, it is not just the representation of digital and hypertextual media that is central to their storytelling: critical to the work is the unremitting analogue form of the printed novel itself. It is this that has allowed McIntosh to defamiliarise the mediatised nature of contemporary life, in a kind of textual trompé-l'œil. In the postdigital age, then, the decision to publish in the non-digital domain is increasingly a strategic one.

This switch in dominant/subordinate positions not only impacts how we understand the creative potentiality of the printed book, but also the hidden dynamics of postdigital hybridity itself. Indeed, it is how these tactics, values and conventions – the fundamental poetics of postdigitality – play out more broadly across contemporary storytelling that becomes the focus for the second substantive section of this book.

Notes

1 See https://dreamingmethods.com/miriam.
2 See http://unknownhypertext.com.
3 See www.ineradicablestain.com/dollgames.
4 See www.sunshine69.com.
5 See https://fivedials.com/experiments/#cover.
6 See www.forensic-architecture.org.
7 See https://saydnaya.amnesty.org.
8 I'll be looking at social media in Chapter 7.
9 See www.pottermore.com.
10 See www.panmacmillan.com/blogs/literary/www-thekills-co-uk.

Works cited

Abba, T. and Speakman, D. *These Pages Fall Like Ash*. Bristol: Circumstance, 2013.
Aquilina, M. 'Electronic Literature and the Poetics of Contiguity. *The Bloomsbury Handbook of Electronic Literature*. Ed, Tabbi, J. London: Bloomsbury, 2018, 201–215.
Arellano, R. *Sunshine '69*, 1996. Web. 12 June 2018. www.sunshine69.com.

Ballard, J. G. 'Answers to a Questionnaire' (1985). *J. G. Ballard: The Complete Short Stories*, Volume 2. Ed, Ballard, J. G. London: Fourth Estate, 2011, 657–661.

Barthes, R. 'The Death of the Author'. *Aspen* 5–6, 1967, np.

Baudrillard, J. *Simulations*. Trans, Beitchman, P. et al. New York: Semiotext(e), 1983.

Bedford, M. and Campbell, A. *The Virtual Disappearance of Miriam*, 2000. Web. 9 February 2019. https://dreamingmethods.com/miriam.

Berners-Lee, T. *Weaving the Web: The Original Design and Ultimate Destiny of the World Wide Web by its Inventor*. London: Harper, 1999.

Bolt, B. *Heidegger Reframed*. London: I.B. Tauris, 2011.

Bolter, J. *Writing Space: Computers, Hypertext, and the Remediation of Print*. 2nd Edition. London: Lawrence Erlbaum Associates, 2001.

Bolter, J. and Joyce, M. 'Hypertext and Creative Writing'. *Hypertext '87 Papers*. Chapel Hill: University of North Carolina Press, 1987, 41–50.

Cassidy, L. 'Salford 7/District Six. The Use of Participatory Mapping and Material Artefacts in Cultural Memory Projects'. *Mapping Cultures: Place, Practice, Performance*. Ed Roberts, L. London: Palgrave Macmillan, 2015, 181–198.

Ciccoricco, D. 'Digital Fiction: Networked Narratives'. *Routledge Companion to Experimental Literature*. Eds, Bray, J., Gibbons, A., and McHale, B. London: Routledge, 2012. 469–482.

Coover, R. 'The End of Books'. *New York Times*, 21 June 1992, 11.

Cox, K. 'What Has Made Me? Locating Mother in the Textual Labyrinth of Mark Z. Danielewski's *House of Leaves*', *Critical Survey* 18 (2), 2006, 4–15.

Currie, G. *Narratives and Narrators: A Philosophy of Stories*. Oxford: Oxford University Press, 2010.

Danielewski, M. Z. *House of Leaves*. London: Transworld Publishers, 2000.

Danielewski, M. Z. *The Whalestoe Letters*. New York: Pantheon Books, 2000.

Danius, S. *The Senses of Modernism: Technology, Perception and Aesthetics*. Ithaca, New York: Cornell University Press, 2002.

Davis, L. 'Jury Duty'. *The Collected Stories of Lydia Davis*. London: Penguin, [2001] 2014, 360–372.

de Souza e Silva, A. 'From Cyber to Hybrid: Mobile Technologies as Interfaces of Hybrid Spaces'. *Space and Culture*, 9 (3) 2006: 261–278.

Derrida, J. *Writing and Difference*. Chicago: University of Chicago Press, 1978.

Ensslin, A. *Canonizing Hypertext Explorations and Constructions*. London: Bloomsbury, 2007.

Ensslin, A. and Skains, L. 'Hypertext: Storyspace to Twine'. *The Bloomsbury Handbook of Electronic Literature*. Ed, Tabbi, J. London: Bloomsbury, 2018, 295–309.

Facer, K. and Pahl, K. 'Introduction'. *Valuing Interdisciplinary Collaborative Research: Beyond Impact*. Eds, Facer, K. and Pahl, K. Bristol: Polity Press, 2017. 1–21.

Five Dials. 'The Role of Music in Your Life'. *Five Dials*, 2016. Web. 14 June 2018. https://fivedials.com/experiments/#cover.

Foer, J. S. *Extremely Loud and Incredibly Close*. Boston: Houghton Mifflin, 2005.

Forensic Architecture. *Forensic Architecture*. Web. 19 June 2018. www.forensic-architecture.org/.

Forensic Architecture and Amnesty International. *Saydnaya: Inside a Syrian Torture Prison*. 2016. Web. 25 June 2018. https://saydnaya.amnesty.org.

Gibbons, A. 'Contemporary Autofiction and Metamodern Affect'. *Metamodernism: Historicity, Affect, and Depth After Postmodernism*. Eds, van den Akker, R.

Gibbons, A. and Vermeulen, T. London: Rowman & Littlefield International, 2017, 117–130.

Gibbons, A. *Multimodality, Cognition, and Experimental Literature*. London: Routledge, 2012.

Gillespie, W., Rettberg, S. Stratton, D. and Marquardt, F. *The Unknown*. 1998. Web. 30 April 2018. http://unknownhypertext.com.

Hammond, A. *Literature in the Digital Age: An Introduction*. Cambridge: Cambridge University Press, 2016.

Haraway, D. J. *Staying with the Trouble: Making Kin in the Chthulucene*. Durham: Duke University Press, 2016.

Hawkins, H. *Creativity*. London: Routledge, 2017.

Hayles, N. K. *Electronic Literature: New Horizons for the Literary*. Notre Dame, Indiana: University of Notre Dame Press, 2008.

Holland, M. K. *Succeeding Postmodernism: Language and Humanism in Contemporary American Literature*. London: Bloomsbury, 2013.

House, R. *The Kills*. London: Picador, 2013.

Ingold, T. 'Bindings Against Boundaries: Entanglements of Life in an Open World'. *Environment and Planning A*, 40, 2008: 1796–1810.

Jackson, S. *Patchwork Girl; or a Modern Monster by Mary/Shelley and Herself*. Watertown, MA: Eastgate Systems, 1995.

Jackson, S. and Jackson, P. *The Doll Games*, 2001. Web. 28 May 2018. www.ineradicablestain.com/dollgames.

Johnson, B. S. *The Unfortunates*. London: Panther Books, 1969.

Joyce, M. *afternoon: a story*. Watertown, MA: Eastgate Systems, 1987.

Joyce, M. 'Hypertext and Hypermedia'. *Of Two Minds: Hypertext, Pedagogy and Poetics*. Ann Arbor: University of Michigan Press, 1995 [1993], 19–30.

Joyce, M. *Of Two Minds: Hypertext, Pedagogy and Poetics*. Ann Arbor: University of Michigan Press, 1995.

Kirby, A. *Digimodernism: How New Technologies Dismantle the Postmodern and Reconfigure Our Culture*. London: Continuum, 2009.

Landow, G. P. *Hypertext: The Convergence of Contemporary Critical Theory and Technology*. Baltimore: John Hopkins University, 1992.

Landow, G. P., (ed.) *Hyper/Text/Theory*. Baltimore: John Hopkins University Press, 1994.

Latour, B. *Down to Earth: Politics in the New Climatic Regime*. Cambridge: Polity Press, 2018.

Lipsky, D. *Although of course You End Up Becoming Yourself: A Road Trip with David Foster Wallace*. New York: Broadway Books, 2010.

McIntosh, M. *theMystery.doc*. London: Grove Press, 2017.

Miller, L. 'www.claptrap.com'. *New York Times Book Review*, 15 March 1998. Web. 12 June 2018. www.nytimes.com/books/98/03/15/bookend/bookend.html.

Moore, L. *Self Help*. New York: Alfred. A. Knopf, 1985.

Moulthrop, S. *Victory Garden*. Watertown, MA: Eastgate Systems, 1992.

Phillips, W. and Milner, R. M. *The Ambivalent Internet: Mischief, Oddity and Antagonism Online*. Cambridge: Polity Press, 2017.

Poe, *Haunted*, Atlantic Records, 2000.

Pope, J. 'A Future for Hypertext Fiction'. *The International Journal of Research into New Media Technologies*, 12 (4), 2006: 447–465.

Pope, J. 'The Way Ahead: The Teaching of Hyper-narrative at Bournemouth University'. *New Writing*, 10 (2), 2013: 206–218.

Pope, R. *Creativity: Theory, History, Practice*. London: Routledge, 2005.

Pressman, J. '*House of Leaves*: Reading the Networked Novel'. *Studies in American Fiction*, 34 (1), 2006, 107–128.

Pullinger, K. *Breathe*. Editions at Play, 2018. Web. 5 February 2018. https://editionsatplay.withgoogle.com/#/detail/free-breathe.

Punday, D. 'Narrativity'. *The Bloomsbury Handbook of Electronic Literature*. Ed, Tabbi, J. London: Bloomsbury, 2018, 133–149.

Pynchon. T. *Gravity's Rainbow*. New York: Viking Press, 1973.

Rettberg, S. 'The American Hypertext Novel, and Whatever Became of it?' *Interactive Digital Narrative: History, Theory and Practice*. Eds, Koenitz, H., Ferri, G., Haahr, M., Sezen, D. and Sezen, T. I. London: Routledge, 2015, 24–35.

Rettberg, S. 'Posthyperfiction: Practices in Digital Textuality'. *Interactive Digital Narrative: History, Theory and Practice*. Eds, Koenitz, H., Ferri, G., Haahr, M., Sezen, D. and Sezen, T. I. London: Routledge, 2015, 174–184.

Ryan, J. *A History of the Internet and the Digital Future*. London: Reaktion Books, 2010.

Ryan, M. 'The Interactive Onion: Layers of User Participation in Digital Narrative Texts'. *New Narratives: Stories and Storytelling in the Digital Age*. Eds, Page, R. and Thomas, B. Lincoln: University of Nebraska, 2011, 35–62.

Smyth, K., Power, A. and Martin, R. 'Culturally Mapping Legacies of Collaborative Heritage Projects'. *Valuing Interdisciplinary Collaborative Research: Beyond Impact*. Eds, Facer, K. and Pahl, K. Bristol: Polity Press, 2017, 191–213.

Striphas, T. *The Late Age of Print: Everyday Book Culture from Consumerism to Control*. New York: Columbia University Press, 2009.

Taylor, M. C. *Rewiring the Real: In Conversation with William Gaddis, Richard Powers, Mark Danielewski, and Don DeLillo*. New York: Columbia University Press, 2013.

Thomas, B. 'Trickster Authors and Tricky Readers on the MZD Forums'. *Mark Z. Danielewski*. Eds, Bray, J. and Gibbons, A. Manchester: Manchester University Press, 2011, 86–102.

Tringham, R. 'Creating Narratives of the Past as Recombinant Histories'. *Subjects and Narratives in Archaeology*. Eds, van Dyke, R. M. and Bernbeck, R. Boulder, Colorado: University Press of Colorado, 2015. 27–54.

Walsh, J. *Seed*. London: Editions At Play, 2017. Web. 5 February 2018. https://seed-story.com.

Weizman, E. *Forensic Architecture: Violence at the Threshold of Detectability*. New York: Zone Books, 2017.

Part 2

Into infinity

Towards a postdigital poetics

6 Spatiality and text

Locative mobile storytelling

Introduction

In the summer of 2016, a strange thing began to happen. Across the globe, both adults and children were suddenly seen on the streets, mobile phones thrust out in front of them, eyes glued to the screen, as they wandered chaotically across roads, parks, squares and even into buildings. To the innocent onlooker it was as though these people were quite literally spellbound, held in the grip of some unseen power. Newspapers quickly caught on to what was happening, using words such as 'craze', 'fad' and 'mania'.

The reality was rather less fantastic. If there was any witchcraft involved then it was of the digital kind, summoned by the release of a new smartphone app: *Pokémon Go*. *Pokémon*, of course, is part of the hugely successful international media franchise created by Nintendo Co., Ltd. First released in 1996 for the handheld game console, *Game Boy*, *Pokémon* went on to be developed for the Nintendo console and arcade platforms. *Pokémon Go*, released on 6 July 2016, was the first version of the game to be adapted for the mobile market, however, and was met with almost immediate acclaim. A free download, Guinness World Records had officially awarded the game five records by August 2016 (Swatman 2016), including the most revenue grossed by a mobile game in its first month ($206.5 million), the most downloaded mobile game in its first month (130 million), and the most international charts topped simultaneously for a mobile game in its first month (70 separate countries).

The game itself is relatively straightforward in design: after creating an avatar, players move through a virtual map based on the player's geographical location. The aim is to amass as many experience points (XP) as possible, either by catching Pokémon characters by swiping an on-screen ball at it, fighting at a gym or visiting Pokéstops, all superimposed over a smartphone's live image capture (Figure 6.1).

The success of *Pokémon Go* came with all sorts of claims about its wider effect on society. Most of these were seen to be positive. Katherine Isbister, writing in July 2016, noted that the game helped increase people's physical exercise which in turn improved mood and reduced depression. It also brought people together through new forms of social engagement. Yet there were

Figure 6.1 Three screenshots from *Pokémon Go*.
Source: Dalton White/YouTube. Creative Commons.

problems too and by October of the same year, just three months after its initial release, the game was generating a very different set of headlines: 'What went wrong with Pokémon Go?' (Humphery-Jenner 2016). Revenues had crashed, downloads had fallen and player satisfaction was in freefall. In some ways, the game had become a victim of its own success, suffering from repeated server outages. This was not helped by a distributed denial-of-service attack on those very servers by the hacker group, OurMine, on 17 July 2016 (*Economist* 22 July 2016). Yet there were other, more pragmatic, problems too. Players soon discovered that the game rapidly drained both their phone battery and their data allowance. Broadband connectivity could not be relied on, especially outside of large cities, and even then, there could be problems: the game suffered from high-profile glitches which meant it was prone to crashing. *Pokémon Go* had become 'Pokémon Gone' (Humphery-Jenner 2016).

However, it would be wrong to write off *Pokémon Go* as a failure. In 2017 SuperData Research ranked it as the ninth highest grossing mobile app of that year while, in May 2018, it was estimated that the game had generated $104 million in monthly revenue from 147 million monthly active players, its highest since the game's launch. Perhaps more important than its continuing popularity, however, is the impact that *Pokémon Go* had on the public consciousness in those first few weeks of its release. By the end of July 2016 it would not be an exaggeration to say that augmented reality (AR), geolocationary, gaming had gone from the side lines to the mainstream. As we'll see, this had ramifications not only for the gaming world but also for the wider adoption of AR and location-based functionality by digital artists. The speed of *Pokémon Go's* uptake created some interesting social and cultural dilemmas that were without precedent, particularly around the ethics of location-based gaming. Whereas some locations such as museums, parks and retail outlets actively encouraged players, others, such as churches and memorial sites, did not (Carter and Velloso 2016). After official requests, Niantic, the game's designers, removed Pokémon, PokéStops and *Pokémon Go* gyms from the Hiroshima Memorial Park and the Holocaust Museum in Washington, D.C. Signs began to appear at sites across the globe asking visitors to abstain from the game. Some people even found that their home had suddenly become the focus of unwanted attention (Carter and Velloso 2016).

Even though 2016 is not that long ago, in terms of mobile gaming it already feels a lifetime. The increasing power of smartphones and the spread of high bandwidth has transformed the way we interact with mobile technology, from mapping and navigational software, to video capture and chat. Even concepts such as AR and virtual reality are not as far out as they seemed in 2016, let alone in 2007 when the first iPhone was released. The legacy of *Pokémon Go* can be found in the practice of those artists who are at the forefront of exploring these new found geospatial affordances, work such as Kate Pullinger's *Breathe* (2018) and James Attlee's *The Cartographer's Confession* (2018). Both play with notions of fiction and non-fiction, using the technology to explicitly pull the reader, as a situated and embodied entity, into the co-creation of, what Les Roberts calls, the '*textualities* of space, place and mapping' (2015, 18, italics in original).

It is the exploration of such textualities that lies at the heart of this chapter. In some ways, this naturally follows on from the previous two chapters. If geospatial stories draw on aspects of mobile gaming (Ensslin 2014) then so too are they indebted to the sort of hypertext fiction we've already examined. Indeed, *Breathe* and *The Cartographer's Confession* could be seen as just another step in a continuum, stretching back to the early exploratory hypertext of Michael Joyce and Shelley Jackson. From this perspective, it's not so much ideas around digital storytelling itself that have changed as the affordance offered by mobile technology. Although insightful in many ways, such a position does, however, rather downplay the stepwise change that locative mobile storytelling is beginning to offer authors and readers alike. Indeed, from the perspective of our map/rhizome/string-figure modality, I would argue that location-based storytelling is a transformative development, taking the embodied, situated potential of narrative to new, unexplored areas. And, as we shall see, the research potential of such approaches are significant, not only in terms of dynamic new forms of storytelling, but also in regard to how our perception of, and interaction with, physical places can be manipulated. This is especially so for the world's vast and ever-expanding urban spaces. According to the United Nations Population Fund (UNFPA 2011), 2007 was the first time that more than half the world's population was classed as living in cities; they estimate that by mid-century this will have risen to two-thirds.

Population growth, ecological impact and social unrest are just some of the concerns that have made urbanisation one of the world's most pressing and immediate issues (UNFPA 2007). In other words, how we live in city space has never been more important. Critical here is an understanding of how we actually engage as a species with physical space, both at the individual and community level. The approach taken here is perhaps best summarised by Yingjin Zhang who described the city as 'a state of mind, an order of morality, a pattern of attitudes and ritualized behavior, a network of human connections, and a body of customs and traditions inscribed in certain practices and discourses' (1996, 3–4). From this perspective, the city is as much a subjective instantiation, 'a drama in time' to use Patrick Geddes phrase (2010, 6), as it is a physically objective encounter. The city then is both a performance and an ongoing narrative, what Henri Lefebvre famously called 'spatial practice' (1991, 33) by which space is transformed into place, 'an organised world of meaning' constituted in its relation to human subjects (Tuan 2001, 179). What I argue here is that locative mobile storytelling offers new ways of positively engaging with this process.

As we've already seen, technological innovation can often precipitate profound social and cultural change which quickly finds expression in artistic practice. If we are living in the era of the smartphone, then it is only the latest in a long list of innovations that includes the television, cinema, telephone and phonograph. Their arrival witnessed similar anxiety about their impact on both society and established media forms. Yet, as Hammond notes, it is often at these very moments of technological upheaval that artists produce their greatest and most telling work, pushing back the limits of artistic representation in an exploration of new forms and techniques (2016, Chapter 2). It is the contention of this book that we are at such a moment once again. The rise of mobile computing, in particular, from smartphones and ereaders, to iPads and wearables, continues to have a significant impact on how people interact with the wider world. In part at least, the work explored in this chapter can be seen as direct interventions into this new form of reality, offering syntheses of story and physical exploration, textual and bodily hybridity. I have called this condition postdigital in recognition of the fact that we have left the old binary divide between the digital and non-digital behind. Instead, we have entered a period dominated by hybridity and transmedial experimentation.

As I have already argued, in our postdigital condition it is the *digital* condition that is increasingly recognised as the default (dominant) state, a condition so ubiquitous it is largely invisible. The decision to write, publish and read *printed* fiction is progressively becoming a strategic and conscious act in and of itself, with all sorts of attendant socio-cultural dynamics underlying it. As hopefully I've also shown, this transformation is not negating either the critical or creative potential of printed fiction; rather, it has given it a new found imperative, a new way of looking back at the world from an increasingly heterodoxic (subordinate) position. In the postdigital age, then, the printed novel has become 'that which is not digital'; but, as I'll show in this chapter, for many writers, this has been used as a starting point from which to create new forms and modes of textual engagement with the world, in other words, new ways by which the printed novel can remain radical and essential as a form of critical expression. While Matthew McIntosh's *theMystery.doc* (2017) re-renders the forms and tropes of digital media for the printed page, other works such as Ben Lerner's *10:04* (2014) and Zadie Smith's *NW* (2012) focus on, what might be termed, the psychological impact of an increasingly media-saturated existence, what Ben Lerner's autofictional author describes as 'the disconnect between his internal experience and his social self-presentation' (2014, 66). Ben Masters, in his analysis of *NW*, argues that Smith's novel 'attempts an ethical intervention' (2017, 167), a confrontation with, what he calls, the 'insecurity, uncertainty, self-questioning, and self-doubt' of twenty-first century existence (2017, 172). I would argue that this description applies equally to *10:04* and *theMystery.doc*. Indeed, as the narrator of *10:04* states, '[a]rt has to offer something other than stylized despair' (2014, 93). Understanding what that 'something' is, how artistic practice is changing and why, goes to the core of this book. What seems to be clear is that these are works of a metamodernist inflection, signifying in their own way a shift in creative modality from the labyrinth of mirrors to the map/rhizome/string-figure. In each of these novels, the exploration of the situated and embodied nature of the author/narrator is just as important as it is in digital works such as *Breathe* and *The Cartographer's Confession*; however, the physicality of these printed works explicitly undercuts the digital sensorium with, what amounts to, new forms of literary representation: a new analogue language, if you will, by which we may contemplate a postdigital existence.

This chapter therefore marks an important development in our understanding of postdigital storytelling. If, as Eric Prieto notes, 'great works of literature have a performative dimension … they help make possible the emergence and establishment of new kinds of places' (2013, 9), then in what ways is this process influenced by mobile technology? How does it influence the sort of stories we tell ourselves, and how does it affect our understanding and engagement with the wider world? These and other concerns are explored through a focus that lies beyond storytelling *per se*, embracing a wider consideration of practice-based research and impact within the arts and humanities more generally.

At the heart of both this and the following chapter is an understanding of, what I term, the poetics of postdigitality. Although this is an emergent area, I would argue that it is in the field of both geospatial and collaborative storytelling that such transformative changes to the way we write and tell stories is most readily observed. As elsewhere in this book, my primary interest is stories that are told explicitly through digital platforms; yet I also realise that I cannot ignore the postdigital implications of printed and hybridic works. Indeed, I argue here that locative mobile storytelling is inherently hybridic in nature, explicitly using the embodied movement of the reader to complete the narrative. Postdigitality offers a new way to reconceptualise these transformational changes, both in terms of storytelling but also practice-based research, in which the reader becomes an active and participatory generator of knowledge (see Facer and Pahl 2017).

Space, place, setting

The evocation of place is one of the most powerful aspects of creative writing. It is also one of the most important. It is difficult, perhaps impossible, to think of a plot without at the same time having a sense of the place in which those events unfold. In many creative writing primers, the term *setting* is often used in preference to *place*, despite its rather limited implications (for example, see Linda Anderson and Derek Neale *Writing Fiction* 2009). Setting, however, is certainly not the same as place. Setting suggests a passive backdrop, a stage or screen set, perhaps, something singular and particular in scale such as a room, a street or park, in front of which the action will occur.

Place, as we'll see, is something else altogether. In their book, *Writing Fiction: A Guide to Narrative Craft* (2011), Janet Burroway and Elizabeth and Ned Stuckey-French, provide some much needed clarity when they state that '[s]etting helps define a story's dimensions. Setting grounds a story in place' (2011, 164). This grounding is done through the author's use of specific detail in the description of setting, the chairs in the room, for example, or the wooden seat in the park. These particularities of setting work to suggest the wider context in which the action happens. One aspect of this wider context is place.

The relationship between space, place and setting is complex but a necessary starting point for this chapter. This is not just about an understanding of artistic craft and technique; rather, at its core lies a far-more fundamental comprehension of how we interact with space, both individually and collectively. Space is, what J. E. Malpas calls, 'dimensionality' (2007, 23); in other words, a physical and objective materiality that Ryan, Foote and Azaryahu describe as 'location, position, arrangement, distance, direction, orientation, and movement' (2016, 7). Place, on the other hand, is a more alchemical product, the outcome of the interaction between space and human consciousness. The geographer Yi-Fu Tuan describes place as 'whatever stable object catches our attention', a pause in our movement through space (2001, 161). The verb *placemaking* refers to this process, by which 'individuals as well as social groups shape the environment and invest space with meanings' (Ryan, Foote and Azaryahu 2016, 7). Drawing on Heidegger's concept of *Dasein* (being-in-the-world), Malpas argues that human subjectivity itself is constituted through the 'agency and embodied spatiality' of place (2007, 35). In other words, Malpas puts placemaking at the heart of being human, or, what Heidegger termed, 'the worlding of the world' (Bolt 2011, 87), a concept we've already met. Robert T. Tally Jr. calls this *topophrenia*, 'a constant and uneasy "placemindedness" (2019, 1). Yet if topophrenia describes an essential part of being human, then it also embraces 'a profound sense of unease, anxiety, or discontent' (2019, 23).

The implications of all this are profound. First, it establishes the central idea that place, unlike space, is not an objective given, a material backdrop through which we simply pass. Instead, it exists only as an embodied instantiation – Tuan's pause or resting of the eye – a complex interplay of human consciousness and physical space. Second, and following on from the first, it introduces the concept that the instantiation of any particular place is neither immutable or a universal given. If place is open to the vagaries of human behaviour then it too must be open to forms of change and adaptation. And third, creative expression is one means by which such change and adaptation can be generated.

The notion that literary texts are a means by which our understanding of particular places can be studied is not a new one. Literary geographers have long been interested in the relationship between fictionalised place and the 'extra-textual' or 'real' world (see David McLaughlin 2018). For some, this involves a focus on the often intimate relationship between literary practice and place, something that Angharad Saunders calls 'a writer's geography', a concept that necessarily extends beyond 'the spatial references and inferences of their texts' to the physical spaces associated with authorial process (2017, xxvi). For others, however, literary geography implies a far greater emphasis on active readerly participation and the social benefits that such an approach might bring. The 'Mapping the Lakes' project, for instance, was an early attempt at exploring how a digital representation of Samuel Coleridge's walking tour of August 1802 could be integrated with 'reader-generated mappings' of those same routes (Cooper 2015, 44), transforming mapping into a socio-cultural activity, fundamentally situated, embodied and partial (2015, 49). In fact, for Joanna E. Taylor, Christopher E. Donaldson, Ian N. Gregory and James O. Butler, it is this affective imperative that remains literary geography's most important element, driven on by the continuing advances in geospatial and mobile computing:

> The digital map – and particularly the kind of digital deep mapping we have outlined – mediates between the reader and the text in putting forward a visible representation of how reader, writer and text might inhabit the same geographical space. The rapid technological advancements we continue to witness promise to drive such mediations in the future.
> (2018, 16)

I've already introduced this relationship between mapping and storifying. Maps, like novels, are always fictional abstractions of the real world: as Peter Turchi states, '[t]o ask for a map is to say, "Tell me a story"' (2004, 11). Yet Turchi notes that the reverse is also true: 'a story or novel is a kind of map because, like a map, it is not a world, but it evokes one' (2004, 166). Understanding the complexities of this 'evoking' lies at the heart of this chapter. In John Cheever's short story, 'The Swimmer' (1964), the central character, Ned Merrill, decides to 'swim' home. As Cheever tells us, '[t]he only maps and charts [Ned] had to go by were remembered or imaginary but these were clear enough' (1978, 777). Cheever's story foregrounds the often complex and contradictory relationship that humans have to place, a concept that at all times hovers precariously between the objective dimensionality of space and the emotionally-driven stories of our imagination. As Taylor, Donaldson, Gregory and Butler remind us, through digital technology we have now reached a moment when embodied, real-time placemaking and storifying have coalesced, creating new hybridic forms of narrative. In order to explore these developments, I shall turn to one particular type of contemporary space, city space, what Matthew Beaumont and Gregory Dart call 'restless cities', sites of 'endless making and unmaking' in the modern world (2010, x).

City space

If cities are restless, then they are also dynamic, transformative and capricious. And as more and more of us live within their boundaries, so the future of our species will increasingly depend on sustainable and resilient urban cohabitation. The United Nation's *2018 Revision of World Urbanization Prospects* notes that 55 per cent of the world's current population lives in urban areas, a proportion that is expected to increase to 68 per cent by 2050. By 2030, the world is projected to have 43 megacities with more than 10 million inhabitants, most of them in developing regions. In their publication, *Transforming our World: The 2030 Agenda for Sustainable Development* (2015), the United Nations recognised three critical dimensions to sustainable development across the planet: economic, social and environmental. Goal 11 of the agenda is aimed specifically at making the world's cities inclusive, safe, resilient and sustainable by 2030.

As I discussed in Chapter 2, the primary purpose of creativity within academe is to elicit purposeful and affective change. Let's remind ourselves again of my revised definition:

> Creativity is a situated and embodied act, in which, through the (co-) creation of new knowledge, perceptions and understanding of the world are changed.

It would be difficult to find a theme or area that's more suited to this definition than the research imperatives around urban living. I would argue that the sort of placemaking discussed in this chapter offers an important methodology by which the social dimension of sustainability can be furthered. Indeed, the Connected Communities Programme (Research Councils UK and the Arts and Humanities Research Council) has already highlighted the central role that participatory storytelling can play in addressing these issues through community-led placemaking activities (Facer and Pahl 2017).

Historically, cities have always been foci of social and cultural anxiety. Migration, over-population and destitution would have been nothing new to citizens of nineteenth-century London, New York or Paris; nor would the increasing concern of today's national governments about the economic and political consequences of urbanisation. What has changed, however, is the scale and spread of urbanity. As we've seen, the order of magnitude of our conurbations has fundamentally shifted towards the mega or super city while initiatives such as the United Nations' *Transforming our World* agenda underlines urbanisation as a global imperative. The ecological, social and economic sustainability of our cities is recognised as being key to any wider understanding of the Anthropocene as a period of significant disjuncture, associated with what Clive Hamilton, Christophe Bonneuil and François Gemenne identify as 'transformations of the landscape, urbanisation, species extinctions, resource extraction and waste dumping' (2015, 3).

In this chapter I will explore the interrelationship between digital storifying and placemaking as a means by which these broader imperatives can be addressed; in other words, the chapter is looking at postdigital placemaking as a specific form of creative praxis within city space. As James Atlee, author of *The Cartographer's Confession*, states in an interview for the literary journal *Hotel*, 'I like that idea of writing on a map. The three-way connection between author, reader and place provided by the smartphone's Global Positioning System (GPS) capabilities is certainly key to this piece'. It is the creative affordances of this 'three-way connection' that underpins the different approaches to placemaking considered in this chapter.

There are significant overlaps here with psychogeography, a rather loose term that has come to refer to a literary genre all of its own: what is perhaps best described as autobiographical journeying from as diverse a range of authors as Peter Ackroyd, Will Self, Iain Sinclair and W. G. Sebald (see Coverley 2010). Iain Sinclair, for example, opens *Lights Out for the Territory* (1997) with preparations for a journey in what becomes a sardonic take on the classic adventure tale:

> The notion was to cut a crude V into the sprawl of the city, to vandalise dormant energies by an act of ambulant signmaking … (I had developed this curious conceit while working on my novel *Radon Daughters*: that the physical movements of the characters … might spell out the letters of a secret alphabet …).
>
> (1997, 1)

This emphasis on 'journey' as an embodied autofictional encounter makes psychogeography a useful starting point for our analysis of artistic engagement with space. Here's Robert Macfarlane describing the strange playfulness that lies at psychogeography's heart:

> Unfold a street map of London, place a glass, rim down, anywhere on the map, and draw round its edge. Pick up the map, go out into the city, and walk the circle, keeping as close as you can to the curve. Record the experience as you go, in whatever medium you favour: film, photograph, manuscript, tape. Catch the textual run-off of the streets: the graffiti, the branded litter, the snatches of conversation. Cut for sign. Log the data-stream.
>
> (2005, 3)

Coverley is surely right when he calls contemporary psychogeography a 'meeting point' of concepts and ideas, each with separate, though interconnected, histories and traditions (2010, 11). Yet despite this undoubted fuzziness in its meaning, there still remains a central methodological core that can be traced back to psychogeography's beginnings in postwar Paris and the Lettrist Group, a forerunner of what would become the Situationist International. The situationists drew explicitly on both dadaism and surrealism: as Simon Sadler explains, they looked back nostalgically to a time when 'artists, architects, and designers had pursued disparate, open-ended experiments; for a time when the conditions of modern life – above all, the relationship between "man and machine" – had been addressed head on' (1999, 5). For situationists such as their leader, Guy Debord, these radical humanist traditions of modernism, as they saw it, had been replaced by capitalist orthodoxy. One of the cornerstones of this trend was the publication of the Athens Charter in 1933, a global manifesto produced by the official body of modernist architecture, the Congrès Internationaux d'Architecture Moderne (CIAM). The charter championed, what it termed, 'the functional city' in which rationalist zoning and traffic flows were central elements. As these principles were absorbed into the redevelopment of Europe's battered postwar landscape, Debord recognised two fundamental principles: that this orthodoxy needed to be challenged; and that the site of this resistance should be the city.

Debord believed that one only needed to peel away this functionalist surface to reveal 'the authentic life of the city teeming underneath' (Sadler 1999, 15). Psychogeography was the term Debord used to describe the study of how the geographical environment affected the emotion and behaviour of individuals; with this understanding would then come insight into how such affect could be resisted. Central to psychogeography were two methodologies. The first was the notion of the *dérive*, described by Debord as 'a technique of rapid passage through varied ambiences', in other words, an open, aimless drift through the city, self-consciously transgressing functional design and architecture (Debord 1981, 62). As McKensie Wark notes, the word has aquatic connotations, suggesting 'a space and time of liquid movement, sometimes predictable but sometimes turbulent' (2011, 22). The second was *détournement*, an avant-garde artistic practice, in which the images, signs and texts of the city are reappropriated into new and subversive forms. Through these two approaches, psychogeography became, what Wark calls, 'a *practice* of the city as at once an objective and subjective space' (2011, 27).

One of the finest examples of this is *The Naked City* (1957), a screenprinted map produced by Guy Debord with Asger Jorn. *The Naked City* can be considered an example of *détournement* because of the way it was created out of an official map of Paris: the *Guide Taride de Paris*. Consisting of nineteen fragments interlinked by arrows, Debord and Jorn only included what they saw as the most important districts in terms of the city's psychogeography, in effect, spatially reconfiguring Paris. The arrows offered the reader a guide to possible *dérives* across the city, representing, what Sadler describes as, 'an urban navigational system that operated independently of Paris's dominant patterns of circulation' (1999, 88). It was therefore a map for getting lost, a kind of AR that in its own way was designed to help the reader both see and experience Paris in radical new ways.

Technology was always an aspect of situationist thinking. The issue for Debord was how it could be used innovatively, in ways that undermined its essentialist role within the functional city. Technology might be used as a means of consumerist control but he believed it could also be turned against itself, in playful acts of avant-garde intervention. Perhaps, most famously, was Debord's use of film in his *La Société du Spectacle* (Society of the Spectacle) produced in 1973. Drawing on the concept of détournement, the film was a montage of TV and film excerpts, commercial and industrial film, advertisements and news footage. An even earlier example came in 1959 when the Situationist International made plans to build a labyrinth in Amsterdam's Stedelijk Museum within which visitors would be immersed in, what Sadler calls, a 'psychogeographic assault course' (1999, 115) that included multimedia through the use of recorded sound. The most ambitious part of the plan was the way in which the labyrinth was to be extended outwards to include the entire city: the 'micro-drift organized in this concentrated labyrinth would have to correspond to the operation of the drift that traversed Amsterdam' ('Die Welt als Labyrinth' 1960, 6). Two groups were to be established, each containing three situationists. The groups would spend three days traversing the centre of Amsterdam, their movements dictated by a director of the dérive who would remain in contact through walkie-talkie, 'preparing experiments at certain locations and secretly arranged events' ('Die Welt als Labyrinth' 1960, 6). Unfortunately, however, these intricate plans were cancelled before any actual performance could be undertaken.

Concurrently, the visual artist and situationist, Constant Nieuwenhuys, was working on his masterwork, 'New Babylon'. Known initially as *Dériville* or 'drift city', the project eventually came to consist of numerous maquettes, collages, models, drawings, graphics and texts in which Constant's ideas were set out. From these materials, 'New Babylon' emerged as a utopian city based on the central notion of the dérive and the playful interaction with space (Sadler 1999, 132–138). The disorientation of, what Constant called, a 'dynamic labyrinth' was to be a key feature, a changing, timeless, decentred structure in which individuals creatively interacted with space (1999, 146). What's particularly fascinating about Constant's 'dynamic labyrinth' is the way it anticipates the world wide web. From New Babylon's description as a 'world wide city for the future' (1999, 147), to its instantiation of playful, interactive, virtual space, it is clear that the situationists were reaching out towards a form of spatial exploration that was innovative, exploratory and interdisciplinary.

The late 1950s and early 60s were to be the highpoint of the movement, however: by the late 60s the situationists were largely forgotten. Yet while Debord and his followers remain in the shadows, some of their ideas have not been quite so retiring. Since the 1990s, psychogeography has witnessed a slow revival, not so much as a physical act in and of itself, but rather as something experienced vicariously through published novels and essays.

In fact it would be possible to argue that psychogeography exists today as a separate literary genre all of its own. I've already listed a number of authors that one would immediately put into this genre; one could also easily include a rich variety of historical dérives such as Charlotte Higgins' *Under Another Sky: Journeys in Roman Britain* (2014) in which the author sets out in order to see how 'the idea of Roman Britain has resonated in British culture and still forms part of the texture of its landscape' (2014, xx). In fact, this interest in the textual representation of a contemporary journey across a historicised landscape has turned into an avalanche. In 2017 alone, one could list John Higgs' *Watling Street: Travels through Britain and its Ever-Present Past*, Tom Chesshyre's *From Source to Sea: Notes from a 215-mile Walk Along the River Thames* and James Canton's *Ancient Wonderings: Journeys Into Prehistoric Britain*. Ian Sansom suggests that 'it now seems to be a rite of passage for the middle-class, middle-aged Englishman to go off on a long walk and then to write a book about it' (2017, 32). Yet this gendered bias is overstated. Lauren Elkin has argued that 'walking the city' was not just a 'masculine privilege' (2016, 19), identifying a range of women flâneuses, including Virginia Woolf, George Sand and Jean Rhys, for whom urban perambulation was a significant aspect of their creative practice. Indeed, Elkin's book is itself a psychogeographic work, threading together the past and the present in a deeply autobiographical account. In *Wanderlust: A History of Walking* (1999), Rebecca Solnit describes how walking creates 'an odd consonance between internal and external passage, one that suggests that the mind is also a landscape of sorts and that walking it is one way to traverse it' (6). For Solnit, walking enhances the symbiotic relationship between the physical and emotional that is at the heart of modern-day psychogeographic writing. Indeed, for Solnit, it is the emotional and subjective that is the real arena of exploration, the physical passage providing the means by which the internal landscape is traversed.

All this feels very much in keeping with the general philosophy of psychogeography that Debord might have recognised, in particular the acute sensitivity between a physical and subjective journey. Yet it's also clear that some things are not the same, that this latest incarnation is more of a psychogeography 2.0, if you will, than simply a straightforward continuation of situationist philosophy. With this new iteration, gone is the overt politicisation of the dérive and, with it, a focus on the industrial cities of western capitalism. In its place has come a much stronger interest in the forgotten and overlooked historical legacies of place, contrasting 'a horizontal movement across the topography of the city with a vertical descent through its past' (Coverley 2010, 14). This is not to say that the situationists were not interested in history; as Ivan Chtcheglov stated (before he was thrown out of the group in 1954), '[a]ll cities are geological' (1981, 2). Yet, for the situationists, history was viewed from a Marxist perspective in which a pre-capitalist past was idealised as superior and authentic in ways that the contemporary city could never be. In psychogeography 2.0, however, the landscape has become 'hyper-archaeologicalised', redolent with undiscovered artefacts and historical latency, at the same time that it has also been unanchored from political determinism.[1] From Laura Elkin's 'flâneuse-ing', Iain Sinclair's excavations into London's 'secret history' through to W. G. Sebald's haunting journeys along East Anglia's coastal path, one is left with the sense of the conflation of time, a slow temporal collapsing between past and present.

If psychogeography 2.0 'hyper-archaeologicalises' city space, then it also foregrounds the embodiedness and situatedness of a dérive as fundamental aspects of human experience. Yet it is a situated embodiedness that is increasingly experienced vicariously, in printed textual accounts of dérives or journeys undertaken by an author. While some of these first-person accounts purport to be entirely factual (for example, Higgins, Higgs and Chesshyre), others tread a more complex route between fact and fiction. Sinclair would be an example of an author whose form of storytelling is self-consciously elliptical, digressive and fragmentary; much of the work of Sebald appears to be factual, even with the inclusion of black and white photographs, but is fundamentally fictional in its premise. Teju Cole's *Open City* (2011), on the other hand, while drawing strongly on Sebald's narrative style, is very much a fictionalised conceit of one man's journey. Zadie Smith's *NW* (2012) is another novel constructed around the narrative effects of psychogeography: Wendy Knepper goes as far to say that *NW* deploys an 'interactive experience of worldly/textual navigation and re-routing ... to prompt its readers to remap known relations to place and explore the contested production of localities in a globalising world' (2013, 116).

In these works, the spatial and temporal disorientation of the characters (either real or fictional) defamiliarise the ordinary world for the reader; by making that world strange and unknown, they allow for overt exploration and a search for meaning, a process that then reconnects the reader to the physicality of space. This interactive role by the reader in the evocation of imaginary spaces is not limited to the psychogeographic novel, of course: as Turchi states (2004, 166), all stories are maps in that regard. Yet, in psychogeographic works, this interrelation between space and narrative, the embodied and the external, is made overt. Psychogeography 2.0 puts placemaking at the heart of the narrative process while its continuing popularity as a genre across both non-fiction *and* fiction amounts to a yearning for, what I would call, an *authentic embodiedness*.

In their discussion of narrative theory and space, Ryan, Foote and Azaryahu describe five basic levels of, what they term, 'narrative space' (2016, 23–25). *Spatial frames* describes the immediate surroundings of the characters, in other words, a spatial perimeter that is individual and localised; *setting* on the other hand is used to refer to the wider socio-historico-geographic context of the story, conforming to the use of the same term by Burroway, Stuckey-French and Stuckey-French that I discussed earlier. *Story space* includes both the spatial frames and the settings of the story; however, it also includes those spaces that are either implied or are referred to directly but are not actually the location of physical action. Conversely, *storyworld* is, what Ryan, Foote and Azaryahu refer to as, 'the story space completed by the reader's imagination' (24). In other words, *storyworld* includes all the spatial information referred to in the book, as well the reader's implicit assumptions that naturally extend and deepen the *story space*. As we've seen, Currie calls this process 'pragmatic inference' (2010, 15). Lastly, *narrative universe* embraces all of the previous levels, but also includes the subjective and emotional domain of the characters, what Ryan, Foote and Azaryahu describe as 'counterfactual worlds' instantiated through dreams, beliefs, fantasies and desires (2016, 25).

Other theorists have used different terms and methodological approaches, of course; indeed, the five levels given here draw heavily on Ryan's own use of Possible World Theory (see Ryan 1991). In essence, however, Ryan, Foote and Azaryahu's taxonomy neatly summarises a powerful way by which narrative space can be understood. Narrative space is, in essence, created through a complex interplay between the author and the reader. As one moves through from *spatial frames*, the degree of abstraction increases until we arrive at the *narrative universe*, a point where space can be both a physical and an entirely subjective entity. Yet I would argue that the concept of *authentic embodiedness* demands a further elaboration of this model. What makes psychogeographic writing so compelling is that the divide between 'narrative space' and 'non-narrative space' (in other words, the real world in which the reader is situated) is either partially or completely collapsed. The narrator's psychogeographic journey self-consciously (strategically) reaches out into the physical spaces of the real, extradiegetic, world. The intimacy of the first-person voice, the de rigueur map and amateur photographs, all add to this effect, whether one is reading non-fiction such as Higgins' *Under Another Sky* or fictional work such as Smith's *NW*.

By 'reaching out', I refer to the entangling of real and 'narrative space' as a direct and intentional consequence of narrative effect. All books are experienced in the real world, of course; in that sense, there is always a textual and bodily interaction, a cognitive processing of the written words. Yet, by 'reaching out', I'm referring to much more than that; rather, I'm addressing the condition where textual work deliberately conflates the textual and extra-textual worlds. Through this process a new kind of hybridic space is created, part textual, part real, where the reader becomes the interconnecting medium. I will call this phenomenon *embodied space* and it forms a sixth layer to Ryan, Foote and Azaryahu's taxonomy. *Embodied space* within this context describes a hybridic form of narrative space that arises from the text but which also has direct and intentional connections to spaces within the actual or real world. *Embodied space* is especially important to psychogeography because of the latter's geographical particularity and emphasis on the personal voyage of discovery. In psychogeography, the journey of the narrator in the textual world is firmly rooted in the geography of the real-world; it is both this geographical verisimilitude, together with the genre's narrative form and emplotment, that encourages the reader to vicariously join the narrator in their journey. *Embodied space* is the result of that co-joining. It is fundamentally hybridic, arising from this deliberate 'reaching out' of textual space into the actual world. However, unlike the five levels outlined by Ryan, Foote and Azaryahu, *embodied space* is not an essential characteristic of a text. It only occurs in certain situations. *Embodied space* is therefore strategic, an intentional narrative effect deployed by the author. *Embodied space* reinforces the situated embodiedness of human experience and underpins what I've already identified as the *authentic embodiedness* of psychogeographical writing.

Much of *NW* is set in Willesden and the characters spend a great deal of time journeying across the locality, metaphorically beating the bounds of the spatial limits of their lives (see Knepper 2013). Chapters nine and ten are placed on facing pages – the former on the verso (exactly a page in length), and the latter on the recto. Chapter nine parodies the instructions offered by Google Maps (Smith 2012, 38). Here's the first half of the chapter:

> From A to B:
> A. Yates Lane, London NW8, UK
> B. Bartlett Avenue, London, NW6, UK
> *Walking directions to Bartlett Avenue, London NW6, UK*
> *Suggested routes*
> *A5*
> *2.4 miles*
> *A5 and Salusbury Rd*
> *2.5 miles*
> *A404/Harrow Road*
> *2.8 miles*

The chapter ends with the declaration: 'These directions are for planning purposes only. You may find that construction projects, traffic, weather, or other events may cause conditions to differ from the map results …' (Smith 2012, 38).

On the verso, however, is something both different and the same (Smith 2012, 39):

> From A to B redux:
>
> Sweet stink of the hookah, couscous, kebab, exhaust fumes of a bus deadlock. 98, 16, 32, standing room only – quicker to walk!

It's the same journey (Yates Lane to Bartlett Avenue) but this time it's described as a redux, in other words, something brought back to life after the functionalism of the mapping software. What follows is over a page of sensory detail, a stream of consciousness in which the embodied sensorium is laid out in front of us in the course of a journey: 'Here is the school where they stabbed the headmaster. Here is the Islamic Centre of England opposite the Queen's Arms' (Smith 2012, 40).

In part, because of these narrative techniques and forms, Knepper calls *NW* an 'interactive and immersive book' (2013, 116), one that 'prompts its readers to remap known locations to place and explore the contested production of localities in a globalizing world' (2013, 116). This notion of interactivity was not lost on the publishers. Alongside the novel, Penguin produced an interactive web page showing a Google map of London as the background image (2012).[2] On top of this is placed four Google place markers, each referring to a key location within the novel: Willesden Lane, 274 Kilburn High Road, Camden Lock and 37 Ridley Avenue. Clicking on these place markers activates a video in which Smith reads the relevant, geospecific, extract. Concurrently, key text is juxtaposed on top of photographs from the location. As Cooper notes in his analysis of this material as a geovisual tool, the web page and video reinforces the interconnectedness of locality (2016, 282–283). As a digital paratext, however, it's one very much geared towards exploring commercial and promotional opportunities. Its use as a storytelling medium in and of itself is limited.

Yet, it's rather too easy to dismiss the *NW* website. After all, it was developed in 2012 when mobile geospatial technology was still very much in its infancy, especially in regard to storytelling. Much has changed since then, especially in regard to urban-based, location-dependent, stories. Indeed, as we'll see, the mobile, geospatial affordance of the average smartphone has revolutionised the opportunities for authors to explore *authentic embodiedness* as a transformative narrative effect. In their study of mobile media, Sarah Pink and Larissa Hjorth note how such locative technologies call forth new forms of, what they term, 'online/offline' entanglements (2014, 496). In other words, the technologically-mediated landscape of embodied space is but one important characteristic of what has remained the primary focus of this book: postdigitality.

Cities and technology

Cities have always been intimately bound up with technological change. As crucial nodes within commercial and communication networks, cities were the centres of sweeping industrialisation that affected all facets of life from the mid nineteenth century (Mumford 1973). In many cases the very form of the city went through profound upheaval as economic function changed and populations swelled, spurred on by the growth of the railways (Morris and Rodger 1993). By the early twentieth century, cities were unrecognisable from their nineteenth-century antecedents, thanks to the proliferation of vast new suburbs and the development of the motor car. From the subterranean sewers, underground tunnels, water mains, electric cable and telephone lines, to the systematised running of roads, rail, housing estates and shopping centres, cities were being enveloped in ever more sophisticated technological networks. The advent of the microprocessor in the late 1960s had only a secondary effect on urbanisation. However, today it's clear that this is no longer the case: digital technology, whether on the high street, at home, in the office or factory, has become a major disruptor. The apotheosis of this is the rise of the 'smart city' (see Townsend 2013; Jordan 2015). Driven in part by policy makers, the term has come to encapsulate a city in which new media help regulate, run and manage many facets of urban living (Caragliu *et al.* 2009). Key here is what has come to be called 'big data', the generation of vast quantities of information with, as Batty notes, 'smartness in cities pertaining primarily to the ways in which sensors can generate new data streams in real time with precise geopositioning …' (2013, 276). Critical to these developments is both the rise of mobile technology and ubiquitous wifi, a phenomenon I've already discussed. As Batty states, 'we are now seeing computers being embedded into every conceivable type of object, including ourselves …' (2013, 275). In the smart city, every car and citizen will become a live node, sending back data in real time by which the city will learn and adapt. For smart city acolytes such as Anthony M. Townsend, the rise of the 'Internet of Things' will only further entrench digital technology as 'an immanent force that pervades and sustains our urban world' (2013, 4). It is also clear, however, that the ethics surrounding this personal data economy are still very much undeveloped.[3]

One of the key challenges facing us today is understanding how such technological innovations can ethically and consensually support urban living in the coming years. Whether we use terms such as smart city, informational city (Castells 1989) or Scott McQuire's 'media city' (2008), sustainable and resilient cities remains a global concern, as I've shown. The social, economic and environmental issues highlighted in the United Nations' publication, *Transforming our World* (2015), provide a powerful imperative for any technological intervention. As I and others have argued, digital storytelling is just one way by which the social and cultural issues surrounding, what might be termed, resilient and sustainable co-existence, can be addressed (Jordan 2015, 2016). Yet, our understanding of the impact of digital technologies on urban living is still in its infancy. As McQuire has noted, 'media cities' provide 'a distinctive mode of social experience' whose 'full implications are only now coming to the fore' (2008, vi).

Historically, of course, alterations to the urban form have always manifested profound changes for those living within the city. Georg Simmel saw the modern nineteenth-century metropolis as being characterised both by strangers and by the experience of 'shock'. And as Jane Jacobs observed, writing in the early 1960s, '...cities are, by definition, full of strangers. To any one person, strangers are far more common in big cities than acquaintances' (2000, 40). Whereas, historically, cities were seen as having a clear centre around which the urban form radiated (Sassen 1991, 13), modern cities have come to be characterised more by their lack of identifiable centres: 'We live ... in an exploding universe of mechanical and electronic invention ... This technological explosion has produced a similar explosion of the city itself: the city has burst open ...' (Mumford 1973, 45). Robert Fishman is even more dogmatic: 'The new city ... lacks what gave shape and meaning to every urban form of the past: a dominant single core and definable boundaries' (1994, 398). The result is a new kind of space, neither city nor countryside, what Franco Ferrarotti calls 'an urban-rural continuum' (1994, 463), a city without a place, 'ageographical' to use Michael Sorkin's description (1992, xi), or Edward Soja's 'postmetropolis' (2000, 95). As Néstor García Canclini wrote at the beginning of this century: 'I end up asking myself if we will be able to narrate the city again. Can there be stories in our cities, dominated as they are by disconnection, atomization, and insignificance' (2001, 85). The rise of psychogeography 2.0 is, at least in part, a response to such existential anxiety.

At the same time that locality, community and neighbourhood have come under threat as meaningful socio-cultural entities, digital technology has emerged as a major urban disruptor. McQuire argues that these conditions have created a new kind of space, what he calls *relational space* (2008, 20–26). McQuire uses the term to describe the impact that digital technology has on our experience of space: 'relational space is the social space created by the contemporary imperative to actively establish social relations "on the fly" ... in which the global is inextricably imbricated with the face-to-face' (2008, 23). One of the tensions within relational space, then, is between this at-distance, online connectedness, and the physical (bodily) estrangement of life within twenty-first century cities. As McQuire notes:

> Suspended between the resurgent promise of technological ubiquity and the recurrent threat of technological alienation, there is an urgent need to investigate what it means today to be 'at home'. Does this still correspond to a particular location, site or territory – or rather, to a particular sense of situation, of locatedness, of cultural belonging?
>
> (2008, 8)

In its hybridic form, relational space shares many of the characteristics of Pink and Hjorth's 'online/offline' entanglements (2014, 496). Both are concerned with the new creative possibilities opened up by the interplay between digital technology and, what I've called, embodied space; as a consequence, both can also be understood as significant features of postdigitality. Yet there's also another imperative at play here. The exploration of these hybridic spaces, with their underlying tension between digital connectivity and physical disconnection, cannot be left to the commercial ambitions of media and technology companies such as Nintendo Co., Ltd., as fun and entertaining as their products might be. If we are to take full advantage of what these transformed, postdigital encounters can offer, then creative artists, practitioners and researchers need to be at the forefront of any developments, opening up radical new ways of placemaking. It is to an examination of such work that this chapter now turns.

Postdigital wayfaring: locative mobile storytelling

Mobile media is nothing new. Portable handheld publications such as chapbooks, pamphlets and newssheets existed for hundreds of years before the mobile phone. The rise of the cheap paperback in the 1930s provided commuters with a vast array of readily-available titles in which to immerse themselves. But then, in 1979, Sony released the Walkman, a clunky box that played compact cassette tape through attachable headphones (Figure 6.2). It was an immediate sensation and helped change the way people interacted with electronic technology. In 2001 Apple released the iPod, a portable 5G hard drive that could store up to a thousand digital songs. Like the Walkman before it, the iPod, with its futuristic design by Sir Jonathan Ive, transformed the mobile media market. Since then, there has been a proliferation of mobile devices, including PDAs, ereaders and iPads, but the real game changer, of course, has been the smartphone.

The smartphone has revolutionised the way we interact with each other and with physical space. The confluence of GPS, 3, 4 and 5G capability, touch screen, gyroscopic and camera technologies, together with improving battery performance and wifi coverage, has meant that digital authors have, in a short space of time, been presented with a radical new creative tool. The combination of location awareness, pervasive wifi connectivity and the availability of products and services through online app stores has meant that smartphones have quickly become extremely powerful, and ubiquitous, locative media (see Jordan Frith 2015, Chapter 3). Indeed, Jason Farman goes as far as to argue that locative media now constitute 'the interface of everyday life' (2012, 87):

> In cities throughout the world … navigating the landscape is a simultaneous process of sensorial movement through streets and buildings … and an embodied connection to how those places are augmented by digital information on mobile devices…The relationship between these interfaces has become so seamless that it has completely altered the way we embody the landscapes we inhabit.
>
> (2012, 87)

Both Farman (2012) and de Souza e Silva (2006) use the term 'hybrid space'; McQuire, as we've seen, prefers the label 'relational space'. Yet, in essence,

Figure 6.2 First generation Sony Walkman TPS-L2 (1979).
Source: Binarysequence, Wikimedia Commons, CC BY-SA 4.0.

they're categorising the same thing: new forms of postdigital spatiality opened up by this pervasive interplay between the physical and digital domains. Crucially, however, neither hybrid nor relational space is a simple outcome of technology: rather, what is central to their construction is the materialising social practices across and between physical and digital spaces (see de Souza e Silva 2006, 265–266).

For some, such as de Souza e Silva, this social practice is conceptualised as a network. Yet this comes with its own implicit set of assumptions, some of which can, and should, be questioned. As Larissa Hjorth and Michael Arnold argue, a network 'is a profoundly cynical model of sociality' (2013, 132) whose key feature is functional (nodal) independence and resilience: '[t]here is therefore no sense of solidarity across the network, no sense of tradition, no common identity … and no common interests' (133). Instead, Hjorth and Arnold propose a new kind of conceptual framework by which phenomena such as hybrid space might be better understood. Rather than network, they propose the term 'intimate and social publics' (9–14), foregrounding the positive impact that sociotechnological changes have had on our experience of personal and collective intimacy, on our fundamental being-in-the-world. As

Hjorth and Arnold state, 'our communications technologies, our interpersonal communication, and our concept and performance of intimacy come together in daily life in a performance that hybridises technology, communication and intimacy' (10). These hybridic, online/offline publics are always emergent, are always in the process of being materialised through the collective practice of its constituents.

I would argue that such 'publics' are better conceptualised as a form of *meshwork*, Tim Ingold's own anthropological repost to network theory (2016). For Ingold, 'the lines of the meshwork are the trails along which life is lived ... it is the entanglement of lines, not in the connecting of points, that the mesh is constituted' (83). In a meshwork, then, it is our very entanglement, our being-in-the-world, that constitutes the 'inhabited world'. Ingold uses the term 'wayfaring' to describe this movement 'along'. Although Ingold was not explicitly referring to online interaction, Pink and Hjorth recognise that these ideas can be used to model our online/offline entanglement: they adopt the term 'digital wayfaring' to describe this movement between digital and physical domains (2014, 491). I would suggest, however, that 'postdigital wayfaring' is a more apposite term here, given what I've already laid out in this book. Yet, whatever term is preferred, the essential point is that wayfaring emphasises the affective, embodied and situated nature of our entanglement with hybridic space. As Ingold states:

> Far from connecting points in a network, every relation is one line in a meshwork of interwoven trails. To tell a story, then, is to *relate*, in narrative, the occurrences of the past, retracing a path through the world that others, recursively picking up the threads of past lives, can follow in the process of spinning out their own.
>
> (2016, 93)

From this perspective, the smartphone becomes a kind of 'intelligent compass', as Martijn de Waal calls it, and digital works such as *The Cartographer's Confession* (2018), creative interventions into postdigital wayfaring. With *The Cartographer's Confession*, James Attlee has created a locative mobile story set in London. Combining fiction, non-fiction, audio, video, music, maps and photographs, the story tells the tale of a young Norwegian boy, Thomas Andersen, and his mother who come to London as refugees at the end of the Second World War. In 1951 Thomas' mother disappears and much of what constitutes *The Cartographer's Confession* is an investigation into what actually happened. Yet the story is told elliptically through a broken trail of textual, audio and photographic materials. It is also designed as a sequence of nested narratives, each operating at a different time, but, crucially, sharing the same discrete set of locations in central London. The opening narrative is set 'today' in which we are told about the discovery of, the now deceased, Thomas Andersen's archive of documents and audio cassettes by the screenwriter, Catriona Schilling. Schilling realises that buried away in these records is some kind of clue to Andersen's life. Nested within this narrative are the audio-cassettes of Andersen himself, recorded in the 1960s when he was looking back on his life. Alongside this audio record are photographs and letters from the 1940s. And finally, around all three of these narratives, is that of the reader themselves, an embodied, interactive, narrative in which they bring the diverse elements of the story together through their own physical exploration of the city. As the author, James Atlee, has stated in an essay for *Review 31*:

> There is no denying the electric charge generated when setting and narrative combine and you feel located within the story itself: at such moments it feels as if the characters whose voices you are listening to might appear round a corner at any moment. With remarkable synchronicity, a bird cries overhead at the moment it is mentioned in the text; you look up, to discover it exists only in the 3D soundscape that surrounds you. Sitting on the top deck of a bus with an extended mix of the theme music still playing in your headphones, you scroll your phone for messages when you are interrupted by an alert that another piece of the story has arrived: writing is reaching you where you read and where you are. All of this potential has to be exciting to any author.

The Cartographer's Confession does at least two things: the first, and the most conventional, is that it reminds us of the historical and social complexity of any space, that etched on the streets of every city are countless forgotten tragedies and vicissitudes. As migrants, Andersen and his mother immediately resonate with any twenty-first century reader, their status and the tragic circumstances of their arrival operating both as an historical record but also an emotional marker within a centuries-long continuum of social, economic and military displacement. Yet secondly, *The Cartographer's Confession* uses the capabilities of the smartphone to bring the reader physically into the story. This spatial entanglement of reader and story, whether it be outside the Tate Modern or Borough Market, pulls the reader back through time. In this way, the story defamiliarises both these locations and the relationship that the reader has with such spaces. What starts off as an investigation into the life of a Second World War migrant, ends up as a dissection of the reader's own understanding of the relationship between contemporary space and memory.

The Cartographer's Confession fits neatly into my revised typology of, what Ryan, Foote and Azaryahu, call 'narrative space' (2016, 23–25). Spatial frames, setting, story space, storyworld and narrative universe all clearly exist across and between the fractured, multi-layered story that Atlee has produced. Yet Thomas Andersen and Catriona Schilling are not the only primary characters in the story; the third major character is the reader, and this creates the sixth and final 'narrative space' that I have described, namely, *embodied space*. In *The Cartographer's Confession*, the reader is active cognitively and physically, exploring both the narrative space of the text while, concurrently, exploring those same physical spaces in the real world. As I've shown, for psychogeography 2.0, a genre dominated by printed-based books, that physical exploration is still very much a vicarious pleasure, a cognitive connection, summoned up by the reader as they sit in their armchair. Yet, by taking advantage of the smartphone, authors such as Atlee are exploring the creative potential of a new kind of storytelling, a storytelling that is both locative and mobile. *Embodied space* is the narrative space created by this innovative set of affordances. It is hybridic in the sense that it is instantiated through the direct and strategic interplay between the narrative space of the text and the physical space of the reader. In the case of *The Cartographer's Confession*, the smartphone allows the author to lead the reader through real-life settings, triggering site-specific content along the way. The story evolves not simply through the function of reading, but also through the physical journeying, or wayfaring, of the reader.

Facer and Pahl, in their study of the UK Connected Communities Programme, have noted the opportunity that both *embodied learning* – in other words, learning that is fundamentally embodied and situated – and *stories* can play in the generation of individual or community-based knowledge (2017, 17). As they note of many of the projects in their study, '[s]tories … became the way in which "thinking differently" was enabled' (2017, 226). Although not an overt objective of *The Cartographer's Confession*, it is clear that embodied space inherently leads to affective, embodied change. This kind of 'thinking differently' is not just in regard to oneself or to an understanding of the story; rather, it involves changing perceptions of the spaces and communities we inhabit, challenging perhaps, perceived notions of heritage and identity. In other words, embodied space becomes the critical site for placemaking. In Chapter 2, I discussed *Preserving Place: A Cultural Mapping Exercise*, a project funded by the AHRC under the Connected Communities Programme (Smyth *et al.* 2017). The concept of 'cultural mapping' offers a useful way of thinking about how postdigital stories such as *The Cartographer's Confession* purposely affect change, positioning them not simply as digital stories, but rather tactical interventions, a hybridic encounter or wayfaring, between author, reader and place.

A different type of story, but one that, nonetheless, is app-based and specifically locative and mobile, is *People's Journeys / Teithiau Pobl.*[4] The project sought to explore the potential of community-based digital storytelling as a means of fostering the embodied learning that Facer and Pahl discuss. To do this the project set out to record individual narratives across the district of Grangetown in Cardiff, electronically tagging them to specific locations. Grangetown is a small area in the centre of Cardiff. It has a diverse population, with an old Welsh-speaking community, as well as significant Somali and Asian populations. Rather like most UK cities, the area experienced significant slum clearance in the late 1960s and early 1970s. The users of the *People's Journeys* app would be able to re-trace these 'journeys' using a mobile phone to engage with audio, image and video, in effect creating their own 'journey' or 'story' as they moved across Grangetown (Jordan 2016).

As an initial iteration, it was decided to start with a series of discrete community-led narratives relating to the First World War. Drawing on various community members, including local history experts, the project team identified and then recorded nine biographical stories, each of a soldier killed in the Great War from the Grangetown area. These recordings were co-scripted and recorded by community members. Images and video were again sourced by the local community, overseen by the project team.

The app itself was built using the (now defunct) *AppFurnace* mobile application development platform. Each narrative was digitally mapped by the app to create the first of what ultimately was conceived as a limitless number of journeys crisscrossing a community. It was decided that the start of the journey should be the war memorial in the local park. The app incorporated a live Twitter feed through which participants could leave textual responses and photographs. Using an experience design framework, the app was trialled by community representatives. Their feedback was then integrated into the final version. The app was finally published on both Google Play and Apple's App Store.

Some of the locations used in the app are buildings associated with the soldiers, most notably their homes. Clive Street is the longest street in

Figure 6.3 Screenshot from *People's Journey* app (2014).
Source: © Author.

Grangetown and at least thirty-nine soldiers and seamen who lived there died in the First World War, including Private Tom Goodland whose story is included in *People's Journeys*; certain locations have been demolished such as Oakley Street where Rifleman Joseph Taylor was born. Yet, through the use of photographs and community memory, the app attempted to bring such places back into existence.

In the Streets Museum project, Salford, the physical installations travelled around the city, forming a mobile museum which 'does not rely on visitors attending a fixed museum space' (Cassidy 2015, 192). Cassidy notes how such installations encouraged a form of 'social practice' in which the project became 'a vehicle for recalling people's individual and collective pasts and reforming social networks' (2015, 192). Although the Streets Museum project involved the interaction with physical objects, what Cassidy calls 'material culture' (2015), and *People's Journeys* with the subjective landscape of narrative, both have important overlaps. Each was community-led in its construction. And both projects recognised the importance of having the artefact (installation and app) embedded within that community. Yet, in its embracing of mobile technology, *People's Journeys* was able to utilise the narrative potential of embodied space, the physical and cognitive intermingling of narrative and space. Like *The Cartographer's Confession*, or Kate Pullinger's *Breathe*, *People's Journeys* offers an intervention in what I have called postdigital wayfaring – the embodied movement through our hybridic enmeshment – and in that way is suggestive of new forms of storytelling.

One of the key concepts that emerges from these works is the notion of location. As Rowan Wilken and Gerard Goggin note, the contemporary rise of location as a concept has been driven by locative media, such as the smartphone. Traditionally, of course, location has referred to the geographical position of an object; location was subsidiary to space, place or even setting, through which it was instantiated. Yet, as Wilken and Goggin state, '[m]ore recently, however – and commensurate with the rise of location-enabled mobile communications technologies – location is viewed as having taken on increased (or renewed) conceptual importance in its own right' (2015, 3). As de Souza e Silva and Frith explain, where once location was understood from a primarily empirical perspective, the term has recently taken on, what they term, 'complex, multifaceted identities that expand and shift according to the information ascribed to them' (2012, 10). The embodied space of postdigital stories such as *The Cartographer's Confession* and *People's Journeys* is one way in which location has been reinscribed. Or, put another way, embodied space can be understood as the real-time enmeshment of location, place and space. 'Locative media', then, refers to various mobile and location-aware technologies by which location is reinscribed with meaning. These media, of course, and the underlying GPS, have arisen through both military and commercial development. As Andrea Zeffiro recognises, this creates a tension between the operational ethos of the technology and the creative ambitions of the artists and writers who wish to utilise them (2015).

Zeffiro uses 'locative praxis' to refer to the process by which artistic practice itself can become an overt investigation into these tensions. As she states, '[l]ocative praxis, therefore, is a critical and interpretive framework for analyzing the ways in which experimental locative media might intervene in the spatial politics of a space/place' (2015, 69). In line with Wilken and Goggin, and de Souza e Silva and Frith, Zeffiro understands location as a phenomenon that is both socially *and* technologically situated. And it is this hybridity that makes location and locative praxis such critical concepts for postdigital storytelling. Zeffiro's Transborder Immigrant Tool is designed to provide assistance to migrants crossing the Mexico-United States border, recording the site of water reserves and safe locations. Yet, from the perspective of locative praxis, the project becomes less about the software and narrative design *per se*, and much more about the hybridic interplay between embodied experience and technology. Zeffiro uses the term border poetics to refer to this wider enmeshment. Approaching the Transborder Immigrant Tool as an intervention of such border poetics – what Zeffiro calls, 'a cultural expression of border formations and experiences' (2015, 72) – provides the critical perspective from which to assess the socio-technological impact on counter narrative, dissent and transgression.

Applying the concept of locative praxis to *People's Journey* and *The Cartographer's Confession* is equally explanatory. Indeed, like the Transborder Immigrant Tool, both projects are fundamentally an intervention into the kinds of embodied knowledge making and placemaking that Facer and Pahl highlight in their study. Although on a different physical and emotional scale to the transborder migrations discussed by Zeffiro, both *People's Journey* and *The Cartographer's Confession* involve, what might be called, new forms of transgressive poetics, a strategic breaking down of physical and emotional barriers in how we experience city space, and a reaching out to new forms of postdigital enmeshment.

Such approaches, of course, require storytellers to adopt innovative ideas and concepts. The affordance offered by locative technology is critical here. The reader interacts with a locative mobile story through an *interface*, such as a smartphone. Both the reader and the device are interacting in real time with the physical environment. As Jeff Ritchie notes:

> While the author facilitates this interactivity in designing and developing the interface, the audience helps create the narrative they experience and actually brings the story's discourse into being.
>
> (2013, 56)

An interface is where two separate systems or entities meet: for the author of locative mobile stories, it is where the narrative design of the story directly interacts with the reader. Until there is an embodied interaction with the interface, a meeting of the digital and non-digital domains, there is no story.

Just as we've already seen with hypertext stories, locative mobile narratives have the potential of creating completely open, boundless stories, in which the only constraints are imposed by the reader's options and physical movement. Yet, just as we've also discussed with hypertext stories, there is every chance that this will create frustration and ultimately an abandonment of the narrative by the reader (Pope 2006). In his discussion of locative storytelling, Ritchie uses the term 'narrative value threshold' to refer to the point at which this kind of frustration is engendered: to put it bluntly, it is the point at which the reader's effort is no longer deemed to be worth it in terms of what they're getting back (2013, 58). For the postdigital author, then, 'perceived value' and 'perceived reward' are critical design concepts, far more so than for print-based authors.

In essence, there needs to be some sense of narrative structure, however loosely defined, made available to the reader. And this narrative structure needs to embrace both the physical and the digital domains, connecting the spatial location of the reader with the digital lexia of the story. For this to work, mobile locative storytelling needs to find ways of highlighting specific locations or objects, to alter perspective and to provide navigational guidance, at the very least. Crucial to this is what Ritchie calls 'narrative bridges' (2013, 63). A narrative bridge purposely connects both the digital world of the mobile story with the physical, embodied space of the reader. A bridge can be within the digital story itself, such as a digital map in which the reader's geographical location is indicated (as in the case of *The Cartographer's Confession* and *People's Journeys*), or embedded within the physical environment, such as through a QR code. Either way, the key function of the bridge remains the same, to 'help move the audience's experience of the storyworld between media' (Ritchie 2013, 64), thereby providing a degree of narrative structure through which the reader's experience of the story, and their conceptions of perceived value and perceived reward, are addressed.

Ritchie isolates three types of narrative bridges. Mimetic narrative bridges are designed to be part of the storyworld itself. For example, a character within the story might offer navigational instruction, as in the case of *The Cartographer's Confession* or Kate Pullinger's *Breathe*. Diegetic narrative bridges offer instruction from a position outside the story while disruptive narrative bridges are exactly that, either intentional (self-reflexive) or unintentional disruptions of the storyworld, as in the use of QR codes, a technological bridge that, of and in itself, has no meaning within the story.

These bridges are ultimately a means of control over, what is potentially, a limitless universe where the narrative value threshold can be easily crossed. They also seek to overcome (or bridge) the inherent tensions between the competing narratives of the digital and physical domains. While the mobile platform contains the author's digital story, the physical world is open to all sorts of other influences, events and dramas that can capture, or distract, the reader's attention. A narrative bridge is one way a digital author can exert a degree of control over these issues, negotiating the fine line between exploration and constraint. And clearly they are also an important aspect of postdigital wayfaring itself, in other words part of the strategies and approaches by which we navigate the transmediality of twenty-first-century life.

Duncan Speakman's *It Must Have Been Dark By Then* (2017) explores the interplay between the physical book and locative mobile storytelling. In essence, the work offers a meditation on issues of climate change and depopulation. While the reader moves through a physical space, such as their own city or town, the app delivers music, narration and audio recordings evoking the stories of other communities who are struggling with the effects of climate change, from the Louisiana swamplands to depopulated Latvian villages and the Tunisian Sahara. However, unlike the other locative mobile stories we've looked at, narrative bridging for *It Must Have Been Dark By Then* is performed by two entities: the first is a map, situated within the digital domain; and the second is a printed book within the non-digital, physical domain of the reader. Speakman therefore crafts a story that emerges from the interplay between three elements: the digital story, the printed text and the embodied movement of the reader. This third element is the embodied space that I've already discussed, my sixth layer in Ryan, Foote and Azaryahu's taxonomy: the direct and strategic interplay between the narrative space of the text and the physical space of the reader. Much of the narrative affect is created by the juxtapositions, contrasts and synergies engendered by the strategic interplay between these three elements. In this way *It Must Have Been Dark By Then* sets out to defamiliarise the very places in which the reader is most comfortable: the towns and cities that we might call home.

One of the key ways Speakman achieves this is by disrupting the narrative bridging. The digital map provided by the app, on which the reader's location is plotted, is stripped of any geographical data, as you can see in Figure 6.4. In other words, the map is blank. This makes the role of the digital map more complex than in most locative mobile stories. In the case of *It Must Have Been Dark By Then*, the map becomes a critique of its own form, a question thrown back to the reader about their own understanding of place, their placemindedness, or topophrenia, to use Tally's term (2019, 1). The instructions are provided by a diegetic narrator who, at the very beginning of the story, instructs the reader to simply start walking, to drift. There is no predetermined route; instead, the app constructs a unique map for each reader's journey.

Along with the narrator's voice, there is a sophisticated aural landscape of music and recorded sound. The journey is constructed around a series of tasks that involve the user finding a building in which someone lives, a water source (an oasis) such as a stream or river and a junction or meeting of ways. These tasks provide a physical engagement with the primary themes of ecological depredation and human depopulation. As it says in the printed book, the purpose of the work is '… to travel through places whose printed maps and data might become records of things that no longer exist, through a world being visibly and rapidly reshaped' (2019, 8).

Figure 6.4 It Must Have Been Dark By Then (2017), showing book and app.
Source: © Duncan Speakman.

At various points within the journey, the app prompts the reader to read one of the ten chapters in the book. The book pulls across the interactivity of the app through both the form of the text itself and formatting. For example, a number of the pages are printed on semi-transparent paper which allows for the playful interplay between text on sequential (recto) pages. The outward stage of the journey ends with the user being prompted to find some form of barrier or border, as Zeffiro would define it, through which they cannot, or should not, pass. From here they are invited to return to the starting point, retracing their route from memory. Neither the app nor the book is used for this part of the journey.

I use the word meditation about this work because that is very much how it feels: the various elements of the story are brought together by Speakman in a way that is designed to foster affective change in the reader's understanding of both their own place and its inherent connectedness to global crises. The *terra incognita* of the empty map is both home and not-home, place but also not-place, an infinity of possibility to which each of us are connected.

The hybridic nature of embodied space is made overt through the use of the printed text. It also continues Speakman's interest in the interplay between digital and physical storytelling that we've already seen in *These Pages Fall*

Like Ash (Abba and Speakman 2013). This transmediality captures an inherent aspect of our postdigital condition, of course, one by which we are forever moving between and across digital and non-digital domains. Yet, at the same time, Speakman is deliberately utilising the postdigital condition of the printed text, works which, by their very physical printedness, speak to us explicitly as 'that which is not digital'. *It Must Have Been Dark By Then* is therefore also an exploration of the subordinated role and function of the printed work in our postdigital condition. The enmeshment of digital and non-digital texts within an embodied, physical journey allows Speakman to overtly draw attention to the interplay between the digital interface (smartphone/app in this case), the physical book and the embodied movement of the reader. To stand in a street and read from a smartphone is a socially accepted activity; to stand in that same place and read from a printed book is to transgress those same norms, as anyone who experiences Speakman's work will testify.[5] Speakman deliberately accentuates a printed work's transgressive nature (the condition of it being non-digital) by taking it out onto the street and placing it side by side with its digital equivalent. In *It Must Have Been Dark By Then*, the printed book is thereby also defamiliarised, the work equally a meditation on, and a locative praxis into, how affective embodied storytelling is leveraged across postdigital transmediality as it is on the wider consequences of the Anthropocene.

Postdigital/non-digital wayfaring

As I've already described at the beginning of this chapter, authors of non-digital work such as Zadie Smith and Ben Lerner are already responding to the locative, mobile and hybridic condition of postdigitality. By non-digital, I refer to work that has not been explicitly written for a digital medium. Smith's *NW* (2012) and Lerner's *10:04* (2014) are first and foremost printed novels, works conceived primarily by their authors as hard copy artefacts. Their ereader versions are simply electronic facsimiles of the printed page with no additional functionality.

It is this non-digitality that makes both *NW's* and *10:04's* engagement with digital technology, and in particular, locative mobile storytelling, all the more significant. While *NW* invites the reader to cognitively remap their own relations to (contested) notions of embodied space, *10:04* offers a much more fundamental exploration of, what I would call, metamodernist sensibility (Jordan 2020). Issues of authenticity and meaningful affect are never far from the surface of the novel; indeed, their discussion forms a kind of critical nucleus around which the novel's emplotment is inscribed.

For Lerner, the city (New York, in this case) remains the essential hub, or nodal point, in contemporary life. Just as in *NW*, or Teju Cole's *Open City*, the characters in *10:04* spend a lot of time moving through the city, and these journeys – the street names and buildings – are given in precise detail.

> We walked south among the dimly gleaming disused rails and carefully placed stands of sumac and smoke bush until we reached that part of the High Line where a cut has been made into the deck and wooden steps descend several layers below the structure; the lowest level is fitted with upright windows overlooking Tenth Avenue to form a kind of amphitheater where you can sit and watch the traffic.
>
> (Lerner 2014, 3)

More overtly than these other works, however, Lerner represents the city as the point at which the invisible forces of, what might loosely be termed, global capitalism are more readily experienced, the contradictions of neoliberalism most keenly felt and engaged with. To be in a city is to be on the front line of humanity's striving for meaningful survival in the accumulating crises of the Anthropocene. In *10:04*, the city, like capitalism itself, has become hollowed out, an outmoded signifier for an age that is already past. Lerner uses the adjective 'totaled' to describe this redundancy, a condition that applies equally to the lives of individual women and men. This wider socio-cultural enmeshment, this being-in-the-world, is what Lerner is referring to when he uses the term 'totaled prosody': the cultural logic or ontology of neoliberal postmodernity. The formation of new ways of living, of new post-postmodern prosodies for the twenty-first century, is, at least in part, the responsibility of artists and writers such as Lerner, working their 'way from irony to sincerity in the sinking city, a would-be Whitman of the vulnerable grid' (Lerner 2014, 4).

I would argue there are two aspects to Lerner's 'prosody' that should concern us here. The first is his creative experimentation with autofiction. Unlike Smith's *NW*, *10:04* appears to be autobiographical, about a writer/poet struggling to complete a project. In fact, the process of writing, the movement from an embodied conscious self to its textual representation, is a core subject of the work. In *10:04*, an author, Ben Lerner, is writing, amongst other things, a short story, 'The Golden Vanity', that is about an author, Ben Lerner, who is writing a novel, that turns out to be *10:04*. 'The Golden Vanity' is a real short story, and in real life was indeed published in the *New Yorker*. Yet the story is also included in *10:04* as Chapter 2. In this way it offers a kind of textual transgression, both breaking and redefining the novel's own narrative frame. As we've already seen in Chapter 2, autofiction is a genre in which a text blends autobiography and fiction as a deliberate creative technique (Gibbons 2017). In the past this might have been associated with the sort of postmodern experimentation that came under the rubric of metafiction, in other words fiction that represented knowledge as inherently contrived and relativist. Yet, with the waning of postmodernism, these ontological bounds have slipped; instead metafiction has gradually become post-ironic, or, to use Lee Konstantinou's term, credulous: '[c]redulous metafiction uses metafiction not to cultivate incredulity or irony but rather to foster faith, conviction, immersion and emotional connection' (2017, 93). Contemporary autofiction, then, rather than denying the possibility of an objective self, seeks to reinstantiate it through an interrogation of our being-in-the-world. In other words, the focus of works such as *10:04* is as much the process, as the subject, of being: a renewed engagement with a hermeneutics of the self. *10:04* purposely locates the narrator 'in a place, a time and a body' (Gibbons 2017, 118), a calculated affectivity that seeks an emotional and ethical connection with its readers.

The second aspect of Lerner's work that I want to highlight is his direct representation of postdigitality. In *10:04*, as in real life, digital technology is all pervasive and utterly normalised. In a sense, digitality in *10:04* is everywhere but also nowhere. The smartphone in particular is used incessantly across the novel, as a means of communication, a timepiece and a direction finder. The narrator's journeying across New York is an exploration of these new forms of hybridic spaces:

> So much of the most important personal news I'd received in the last several years had come to me by smartphone while I was abroad in the city that I could plot on a map, could represent spatially, the major events … of my early thirties.
>
> (2014, 32–33)

This postdigital condition induces a kind of sublimated anxiety within *10:04*'s narrator, a subconscious concern about its impact on how we experience the world. The narrator's condition of Marfan, a genetic disorder whose symptoms include, so we're told, 'poor proprioception … a terrible neurological autonomy not only spatial but temporal' (2014, 6–7), can be understood as a direct transference of his own anxieties concerning the technologically-mediated displacement, or cognitive dislocation, induced by his engagement with hybridic space. By asking what remains of traditional notions of place and embodiedness in the hybridic spaces of a postdigital city, *10:04* interrogates the very things that Gibbons isolates as key characteristics of contemporary autofiction: situatedness and embodied subjectivity. The author's anxiety around his proprioception, his failing spatial and temporal awareness, is a constant theme in the novel. Yet, concurrently, the narrator is also hyper sensitised to spatial and temporal affect. Even a gaslight produces a kind of temporal vertigo: 'it was as if the little flame in the gas lamp he paused before were burning at once in the present and in various pasts, in 2012 but also 1912 or 1883, as if it were one flame flickering simultaneously in each of those times, connecting them' (Lerner 2014, 67).

Yet, if the first characteristic of postdigitality in *10:04* is the representation of digitality within the novel itself – from the position of 'that which is not digital' – the second is its tactical emphasis on the value of non-digital form in, what Lerner calls, 'the postcodex world' (2014, 154). *10:04* remains purposely 'old-fashioned' in its emphasis on small presses and the physical purity of hard-copy publication. Lerner has form here: both his collaborative projects, *The Polish Rider* (2018) with Anna Ostoya, and *Blossom* (2015) with

Thomas Demand, are published through Mack, a small press specialising in print-based publications that explore the creative synergy between text and non-textual (photographic) art. *Blossom* (2015), for example, consists of a poem by Lerner, alongside Demand's high-resolution colour photographs of cherry blossom. Although it isn't stated, the tree is actually made of paper, the photographs a subtle examination of fraudulence and authenticity, themes picked up by Lerner's poem. The photographs are spread across both sides of a French fold, sometimes with additional images hiding (and not easily visible) inside. As a result, the book encourages a form of interactivity that we've already seen with the printed book accompanying Speakman's *It Must Have Been Dark by Then*. Both encourage a physical exploration of the printed page in which embodied entanglement is explicitly drawn into the storytelling process.

These themes are equally apparent in *10:04*. Much of the book revolves around the materiality of publication, from the narrator's own work, including a novel we may or may not be reading with its inclusion of the short story, 'The Golden Vanity', to the narrator's self-publication of *To the Future* with Roberto Ortiz. Virtuality as a concept in *10:04* has an unusual duality, referring both to the extrapolated value of any object within the capitalist system (including the narrator's own hypothesised novel) as well as the more common understanding of something being rendered digitally. The effect of this synthesis is to directly equate a digital text with the workings of the market economy, while at the same time positioning the non-digital, or printed text, as something inherently immune from such influence. Yet, crucially, Lerner does not want to wipe away digital technology; instead, rather like Speakman's *It Must Have Been Dark By Then*, *10:04* argues for new forms of creative hybridity, new assemblages of digital and non-digital entanglement. On the very last page Lerner includes a low-grade photograph of Vija Celmins', *Concentric Bearings B*. The work appears to consist of two photographs, one of the night sky and one of a plane. But this is a deliberate illusion, at least on the part of Lerner. The view of the night sky is an aquatint; that of the plane is a mezzotint. Both images were taken from magazine photographs. In the case of the plane, Celmins translated it into a drawing before finally rendering it through the mezzotint process (Rippner 2002). This translation of image – from the physical plane, to photograph, to magazine illustration, to drawing, to mezzotint, before finally back into a digitised photograph in *10:04* – is a beautiful prosody in its own right, at the heart of which is Celmins' own artistic practice, the hand burnishing of a metal plate as part of the mezzotint process. As Lerner says at the very end of *The Polish Rider*, 'the verbal does not get the last word, or gets the last word, but then something else happens: the eloquence of the depiction of silence "talks" back' (2014, 59). In the silence of *Concentric Bearings B* lies the embodied practice of the artist, the physical inscription of handheld tools on a metal plate, before a handmade print can be taken on paper. This prosody of form, technique and non-digital/digital media, is also a prosody of embodied practice and the materiality of creative expression, a new kind of détournement for the metamodernist age that we've also seen in Simon Starling's *ShedBoatShed (Mobile Architecture No. 2)* in Figure 3.3. Yet while the situationists sought new methodologies of transgressive reappropriation within the urban spaces of postwar Europe, postdigital writers such as Lerner, Pullinger, Atlee and Speakman are exploring new forms of hybridity and translation, foregrounding the situated embodiedness, the essential postdigital entanglement, that connects all our lives. As the narrator says at the very end of *10:04*, 'I am with you, and I know how it is' (2014, 240).

Anxiety and apprehension have underlaid much of what I've had to say in this chapter. We've seen it in Ben Lerner's autofictional narrator, with his Marfan and proprioception, and it has also been there in Tally's description of our 'topophrenic condition', a phenomenon characterised by estrangement and unease (2019, 22–23). McQuire's notion of relational space, of the 'deterritorialization of the home', completes this triptych of despair (2008, 10). In all these cases the disruptive impact of digital technology on our experience of intimate space has never been far away; indeed, for McQuire, it is enemy number one.

What I hope this chapter has done is show, in some small way, how locative mobile storytelling can offer purposeful and affective interventions. If the case studies we've looked at are indicative of a new kind of embodied placemaking, then it is a placemaking (Tuan's resting of the eye) not only of our physical spaces, but also the digital sensorium too. In other words, works such as *The Cartographer's Confession* and *It Must Have Been Dark By Then* suggest ways by which we can start to map the relational spaces of our postdigital landscape and, in this way, more meaningfully explore the creative and affective possibilities of the map/rhizome/string-figure.

Summary

The way writers have represented place within their work has always been intimate and complex. And this is especially so when it comes to the city. Early novelists such as Daniel Defoe, Elizabeth Gaskell and Charles Dickens, found news forms of creative agency in the unfolding complexities of urbanisation.[6] The city offered such writers the opportunity to engage dramatically with a wide range of contemporary issues, including social and economic inequality, overcrowding and the wider effects of industrialisation. Yet writers also realised that innovative forms and methods of artistic practice were needed to represent this new world. Modernist works such as James Joyce's *Ulysses* (1922) and Virginia Woolf's *Mrs Dalloway* (1925) can be seen in this light, as can some of the later work I've highlighted under the label of psychogeography, including Iain Sinclair's fusion of poetry and prose, *Lud Heat: A Book of Dead Hamlets* (1975), and more hybridic works such as Zadie Smith's *NW* (2012) and Ben Lerner's *10:04* (2014). From this perspective, then, locative mobile storytelling is simply a continuation of this creative endeavour with works like *The Cartographer's Confession* and *People's Journeys* merely the latest step in how writers have engaged with cityspace.

Yet, I would argue that what we've seen in this chapter suggests that this underplays the innovative impact of locative mobile storytelling. Yes, there is continuity and synthesis with previous work in how narrative and cityspace can be brought together to create a story. The interplay of historical with contemporary space in both *The Cartographer's Confession* and *People's Journeys* draws strongly on psychogeography, of course, but also modernist writing with its emphasis on subjectivity. Yet, I would argue that the use of locative mobile technology has offered writers a stepwise change in how they think about storifying. In short, it represents a revolution in storytelling. Never before have writers been able to create stories that are dependent on the physical movement of the reader, that can consist of a range of media, including audio and video, and that are handheld and mobile. It is the smartphone revolution that has given them that affordance.

Key to this transformation has been the creation of a new kind of narrative space, what I have called *embodied space*. Embodied space is a hybridic form of narrative space that arises from the text but which also has direct and intentional connections to spaces within the actual or real world. I would have argued that it forms a sixth layer in Ryan, Foote and Azaryahu's taxonomy of narrative space. For writers of contemporary psychogeographic works such as Higgins' *Under Another Sky: Journeys in Roman Britain* (2014), embodied space remains an essentially passive phenomenon, experienced by the reader as they sit in their armchair. The reader vicariously experiences the real places described in the work, an immersivity heightened by the narrative's specificities of place and time, often assisted by the inclusion of photographs, maps and sketches. The non-fictional intent of these works complicates the relationship between (fictional) narrative space and real space; indeed, works by authors such as W. G. Sebald deliberately seek to explore this tension. Ultimately, this entanglement between narrative space and real space occurs in the reader's head and remains a cognitive process.

What locative mobile technology has given authors, however, is the ability to explore a far-more dynamic exchange between the textual and physical worlds. In works such as *Breathe* and *The Cartographer's Confession*, embodied space is no longer a passive, cognitive experience but instead, at least in part, a physical journey in the course of which a story evolves. Here, embodied space becomes deeply hybridic, an embodied entanglement of digital and non-digital domains. And as Speakman demonstrates, this hybridity is not necessarily a straightforward one of digital platform and human body. Works such as *It Must Have Been Dark By Then* and *These Pages Fall Like Ash* are early forays towards a more complex form of storytelling, in which an embodied, situated experience is mediated between and across locative mobile technology and physical non-digital texts.

Yet, as an exploration of our 'being-in-the-world', all the stories discussed in this chapter offer, what I term, a form of postdigital wayfaring. This online/

offline wayfaring becomes a way of navigating Ingold's meshwork, what he describes as the 'weave and texture' of the world's 'continual coming into being' (2016, 83). And it is this wayfaring that offers a way of modifying the deleterious effects of McQuire's relational space.

Embodied space, then, is an important aspect of what I have termed a postdigital poetics, as are techniques such as narrative bridging and narrative value threshold. In their ability to embrace affective, situated experiences, they offer the potential of new kinds of storytelling that conform to our revised definition of creativity:

> Creativity is a situated and embodied act, in which, through the (co-) creation of new knowledge, perceptions and understanding of the world are changed.

Edward S. Casey reminds us that 'place serves as the *condition* of all existing things. This means that, far from being merely locatory or situational, place belongs to the very concept of existence' (1993, 15, italics in original). Postdigital wayfaring, then, becomes a fundamental means of engaging with contemporary existence and, ultimately, affecting change in a fragile world. Unlike locative gaming apps such as *Pokémon Go*, works such as *The Cartographer's Confession, People's Journeys* and *It Must Have Been Dark By Then* are not just stories; at the heart of their intent is the desire to induce a change in how we view the world. The same can also be said of non-digital work such as Lerner's *10:04* and Smith's *NW*, print-based novels that explicitly engage with the increasingly hybridic nature of existence.

Central to all these works is the paradigmatic shift that I've already discussed, namely the movement away from the postmodernism of Kearney's labyrinth (2002), towards the adoption of a new set of values that some have grouped under the label, metamodernism. I'm not going to repeat my discussion of metamodernism here; instead I simply note just how neatly locative mobile storytelling fits within metamodernism's critical boundaries. In the hands of writers such as Pullinger, Atlee and Speakman, embodied space becomes a dynamic affordance that foregrounds the situated embodiedness, the essential postdigital entanglement, connecting all our lives. As Patrick Allen notes in his discussion of augmented public spaces, the body is always at the heart of consumption and interpretation, no matter how virtual, or digitally-enhanced the experience (2008, 59). Locative mobile storytelling, then, offers new forms of the sort of metamodernist autofiction that Gibbons has already described. As she states, 'the affective logic of contemporary autofiction is situational in that it narrativises the self, seeking to locate that self in a place, a time and a body. It pertains to represent the truth, however subjective that truth may be' (2017, 118). This description fits equally well as a summary of the kind of locative mobile storytelling I've discussed throughout this chapter as well as broader, postdigital work such as *10:04*. I would argue that the use of embodied space by writers naturally draws any creative work into the orbit of autofiction, with its hybridic mix of fiction and non-fiction, text and body, non-digital and digital. The metamodernist inflection in many mobile locative work is also clear; even in more experimental stories such as *It Must Have Been Dark By Then*, with its denial of geographical specificity, the intent is less about fragmentation and indeterminacy – getting lost in the postmodern labyrinth – and much more about empathy and affective meaning. In Speakman's work, the denial of a map does not signify the denial of the territory, but actually the reverse: the blank screen becomes a void in which the reader is invited to create their own portolan chart in an overt act of embodied placemaking.

These are exciting and innovative ideas with significant implications for storytelling. Yet, as Zeffiro reminds us, the social, cultural and technological implications of locative mobile software should not be forgotten. Indeed, for Zeffiro, locative praxis is at its heart a mode of inquiry into the possibilities and limits of dissent and transgression within and across technological infrastructures. The freedom and utility that comes from precise geographical wayfaring needs to be understood in terms of the fear of being tracked and found, something that remains a particular imperative in work such as the Transborder Immigrant Tool, for example. For Zeffiro, however, critical praxis should not be focussed on the use of such technology *per se*; for her, the possibilities of the technology far outway any overriding cynicism. Rather, attention should be given to how best such technology might be used and, perhaps more importantly, by whom. This suggests a research focus willing to move beyond interdisciplinarity and embrace the sort of transdisciplinarity described by Basarab Nicolescu; in other words, and paraphrasing Nicolescu, that which is between, across and beyond academic disciplines (2008, 2). As Facer and Pahl acknowledge, the use of tacit and embodied knowledge from across diverse communities very often surfaces 'radical ruptures with existing ways of working and modes of knowing' (2017, 227). Embodied space is just one way in which such difference and diversity is manifested. Another is through more collaborative, community-based storytelling. Both this, and its connection to the wider issues of transdisciplinary research within the digital humanities more generally, forms the focus of the next chapter.

Notes

1 There are exceptions. China Miéville's *London's Overthrow* (2012) is an interesting throwback to the more politically-driven imperative of the dérive. See also McKensie Wark's *The Beach Beneath the Street* (2012).
2 See www.penguin.com/static/pages/features/zadie_smith.
3 Cities such as Barcelona, Amsterdam and New York are exploring the ethical use of the personal data economy (see Bass, Sutherland and Symons 2018).
4 The project team included the author and Professor Gareth Loudon (Centre for Creativity Ltd). See https://peoplesjourneys.wordpress.com.
5 Go back ten years and the exact reverse would be true, of course.

6 See Defoe, D. *A Journal of the Plague Year*. London: Penguin Classics, 2003 (1722); Gaskell, E. *North and South*. London: Chapman & Hall, 1855; and Dickens, C. *Oliver Twist; or, The Parish Boy's Progress*. London: Richard Bentley, 1839.

Works cited

Abba, T. and Speakman, D. *These Pages Fall Like Ash*. Bristol: Circumstance, 2013.

Allen, P. 'Framing, Locality and the Body in Augmented Public Space'. *Augmented Urban Spaces: Articulating the Physical and Electronic City*. Eds, Aurigi, A. and De Cindio, F. London: Routledge, 2008, 46–61.

Anderson, L. and Neale, D. *Writing Fiction*. London: Routledge, 2009.

Anon., 'Die Welt als Labyrinth'. *Internationale Situationniste*. 4 June 1960: 5–7.

Atlee, J. 'Take Me to the River: A Journey into Digital Fiction'. *Review 31*. Web. 30 September 2018. http://review31.co.uk/essay/view/53/take-me-to-the-river-a-journey-into-digital-fiction.

Attlee, J. *The Cartographer's Confession*. James Atlee, 2018.

Bass, T. Sutherland, E. and Symons, T. *Reclaiming the Smart City: Personal Data, Trust and the New Commons*. London: Nesta, 2018.

Batty, M. 'Big Data, Smart Cities and City Planning'. *Dialogues in Human Geography*, 3 (3), 2013: 274–279.

Beaumont, M. and Dart, G. 'Preface'. *Restless Cities*. Eds, Beaumont, M. and Dart, G. London: Verso, 2010, ix–xi.

Bolt, B. *Heidegger Reframed*. London: I.B. Tauris, 2011.

Burroway, J., Stuckey-French, E. and Stuckey-French, N. *Writing Fiction: A Guide to Narrative Craft*. London: Pearson, 2011.

Canclini, N. G. *Consumers and Citizens: Globalization and Multicultural Conflicts*. Minneapolis: University of Minnesota Press, 2001.

Canton, J. *Ancient Wonderings: Journeys into Prehistoric Britain*, London: William Collins, 2017

Caragliu, A., Del Bo, C. and Nijkamp, P. 'Smart Cities in Europe'. *Serie Research Memoranda*, (48), 2009, n.p.

Carter, M. and Velloso, E. 'Some Places Should be Off Limits for Games such as Pokémon GO'. *The Conversation*. 12 July 2016. Web. 18 July 2018. https://theconversation.com/some-places-should-be-off-limits-for-games-such-as-pokemon-go-62341.

Casey, E. S. *Getting Back into Place: Toward a Renewed Understanding of the Place-World*. Bloomington: Indiana University Press, 1993.

Cassidy, L. 'Salford 7/ District Six. The Use of Participatory Mapping and Material Artefacts in Cultural Memory Projects'. Mapping Cultures: Place, Practice, Performance. Ed, Roberts, L. London: Palgrave Macmillan, 2015, 181–198.

Castells, M. *The Informational City: Information Technology, Economic Restructuring, and the Urban Regional Process*. Oxford; Cambridge, 1989.

Cheever, J., 'The Swimmer'. *The Stories of John Cheever*. London: Alfred A. Knopf, 1978 [1964], 776–788.

Chesshyre, T. *From Source to Sea: Notes from a 215-mile Walk Along the River Thames*, Chichester: Summersdale Publishers, 2017.

Chtcheglov, I. 'Formulary for a New Urbanism' [1953]. *Situationist International Anthology*. Ed, Knabb, K. Berkeley, CA: Bureau of Public Secrets, 1981, 1–4.

Cole, T. *Open City*. London: Faber and Faber, 2011.

Cooper, D. 'Critical Literary Cartography: Text, Maps and a Coleridge Notebook'. *Mapping Cultures: Place, Practice, Performance*. Ed, Roberts, L. London: Palgrave Macmillan, 2015, 29–52.

Cooper, D. ' "Setting the Globe to Spin": Digital Mapping and Contemporary Literary Culture'. *Literary Mapping in the Digital Age*. Eds, Cooper, D., Donaldson, C. and Murrieta-Flores, P. London: Routledge, 2016, 276–295.

Coverley, M., *Psychogeography*. Harpenden, Hertfordshire: Pocket Essentials, 2010.

Currie, G. *Narratives and Narrators: A Philosophy of Stories*. Oxford: Oxford University Press, 2010.

D'Anastasio, C. 'You Can No Longer Catch Pokémon at Hiroshima's Memorial or The Holocaust Museum'. *Kotaku*, 8 August 2016. Web. 18 July 2018. https://kotaku.com/you-can-no-longer-catch-pokemon-at-hiroshima-1784985508.

Debord, G. 'Theory of the Dérive' [1958]. *Situationist International Anthology*. Ed, Knabb, K. Berkeley, CA: Bureau of Public Secrets, 1981, 62–66.

Demand, T. and Lerner, B. *Blossom*. London: Mack Books, 2015.

The Economist. 'How Pokémon Go was Attacked'. *The Economist*, 22 July (2016). Web. 17 July 2018. www.economist.com/the-economist-explains/2016/07/21/how-pokemon-go-was-attacked.

Elkin, L. *Flâneuse: Women Walk the City in Paris, New York, Tokyo, Venice and London*. London: Vintage, 2016.

Ensslin, A. *Literary Gaming*. Cambridge, Massachusetts: Massachusetts Institute of Technology, 2014.

Facer, K. and Pahl, K., eds. *Valuing Interdisciplinary Collaborative Research: Beyond Impact*. Bristol: Polity Press, 2017.

Farman, J. *Mobile Interface Theory: Embodied Space and Locative Media*. London: Routledge, 2012.

Ferrarotti, F. 'Civil Society as a Polyarchic Form: The City'. *Metropolis: Centre and Symbol of Our Times*. Ed, Kasinitz, P. London: Macmillan, 1994, 450–468.

Fishman, R. 'Megalopolis Unbound'. *Metropolis: Center and Symbol of Our Times*. Ed, Kasinitz, P. London: Macmillan, 1994, 395–412.

Frith, J. *Smartphones as Locative Media*. Cambridge: Polity Press, 2015.

Geddes, P. *Civics: As Applied Sociology*. Boston, MA: Qontro Classic Books (1905), 2010.

Gibbons, A. 'Contemporary Autofiction and Metamodern Affect'. *Metamodernism: Historicity, Affect, and Depth After Postmodernism*. Eds, van den Akker, R. Gibbons, A. and Vermeulen, T. London: Rowman & Littlefield International, 2017, 117–130.

Hamilton, C., Bonneuil, C., and Gemenne, F., 'Thinking the Anthropocene'. *The Anthropocene and the Global Environmental Crisis*. Eds, Hamilton, C., Bonneuil, C. and Gemenne, F. London: Routledge, 2015, 1–13.

Hammond, A. *Literature in the Digital Age: An Introduction*. Cambridge: Cambridge University Press, 2016.

Higgs, J., *Watling Street: Travels through Britain and its Ever-Present Past*, London: Weidenfeld & Nicolson, 2017.

Higgins, C. *Under Another Sky: Journeys in Roman Britain*, London: Vintage, 2014.

Hjorth, L. and Arnold, M. *Online@AsiaPacific: Mobile, Social and Locative Media in the Asia-Pacific*. London: Routledge, 2013.

Hotel. 'Mapping the City, Mapping Chance: James Attlee on The Cartographer's Confession'. *Hotel*. 2018. Web. 30 July 2018. https://partisanhotel.co.uk/James-Attlee.

Humphery-Jenner, M. 'What Went Wrong with *Pokémon Go*? Three Lessons from its Plummeting Player Numbers'. *The Conversation*. 18 October 2016. Web. 17 July 2018. https://theconversation.com/what-went-wrong-with-pokemon-go-three-lessons-from-its-plummeting-player-numbers-67135.

Ingold, T. *Lines: A Brief History*. London: Routledge, 2016.

Isbister, K. 'Why Pokemon Go Became an Instant Phenomenon'. *The Conversation*. 15 July 2016. Web. 17 July 2018. https://theconversation.com/why-pokemon-go-became-an-instant-phenomenon-62412.

Jacobs, J. *The Death and Life of Great American Cities*. London: Pimlico, [1961] 2000.

Jordan, S. 'Hacking the Streets: "Smart" Writing in the Smart City'. *First Monday*, 21 (1), 2016: n.p. Web. 14 August 2018. https://doi.org/10.5210/fm.v21i1.5529.

Jordan, S. 'Street Hauntings: Digital Storytelling in Twenty-First Century Leisure Cultures'. *Digital Leisure Cultures: Critical Perspectives*. Eds, McGillivray, D., Carnicelli, S. and McPherson, G. London: Routledge, 2016, 179–191.

Jordan, S. '"Totaled City": The Postdigital Poetics of Ben Lerner's *10:04*'. *Time, the City, and the Literary Imagination*. Eds, Kramer, K. and Evans, A. London: Palgrave, 2020 (forthcoming).

Jordan, S. 'Writing the Smart City: "Relational Space" and the Concept of "Belonging"'. *Writing in Practice: Journal of Creative Writing Research*, 1 (1), 2015: n.p. Web. 14 August 2018. www.nawe.co.uk/DB/current-wip-edition/articles/writing-the-smart-city-relational-space-and-the-concept-of-belonging.html.

Joyce, J. *Ulysses*. Paris: Sylvia Beach, 1922.

Kearney, R. *On Stories*. London: Routledge, 2002.

Knepper, W. 'Revisionary Modernism and Postmillennial Experimentation in Zadie Smith's *NW*'. *Reading Zadie Smith: The First Decade and Beyond*. Eds, Tew, P. London, Bloomsbury, 2013, 111–126.

Konstantinou, L. 'Four Faces of Postirony'. *Metamodernism: Historicity, Affect, and Depth After Postmodernism*. Eds, van den Akker, R. Gibbons, A. and Vermeulen, T. London: Rowman & Littlefield International, 2017, 87–102.

Lefebvre, H. *The Production of Space*. Oxford: Blackwells, 1991.

Lerner, B. *10:04*. London: Granta, 2014.

Löw, M. (2008) 'The Constitution of Space: The Structuration of Spaces Through the Simultaneity of Effect and Perception'. *European Journal of Social Theory*, 11 (1), 2008: 25–49.

Macfarlane, R. 'A Road of One's Own', *Times Literary Supplement*, 7 October 2005: 3–4.

McIntosh, M. *theMystery.doc*. London: Grove Press, 2017.

McLaughlin, D. 'Thinking (about Literary) Spaces: Ideas from the Cambridge Literary Geographies Conference'. *Literary Geographies* 4 (1), 2018: 1–5.

McQuire, S. *The Media City: Media, Architecture and Urban Space*. London: Sage, 2008.

Malpas, J. E. *Place and Experience: A Philosophical Topography*. Cambridge: Cambridge University Press, 2007.

Masters, B. *Novel Style: Ethics and Excess in English Fiction since the 1960s*. Oxford: Oxford University Press, 2017.

Miéville, C. *London's Overthrow*. London: The Westbourne Press, 2012.

Morris, R. J. and Rodger, R. *The Victorian City: A Reader in British Urban History, 1820–1914*. London: Longman, 1993.

Mumford, L. *The City in History*. London: Harmondsworth, 1973.

Nicolescu, B. '*In Vitro* and *In Vivo* Knowledge – Methodology of Transdisciplinarity'. *Transdisciplinarity: Theory and Practice*. Ed, Nicolescu, B. Cresskill, New Jersey: Hampton Press Inc.: 2008, 1–21.

Ostoya, A. and Lerner, B. *The Polish Rider*. London: Mack Books, 2018.

Pink, S. and Hjorth, L. 'The Digital Wayfarer: Reconceptualizing Camera Phone Practices in an Age of Locative Media'. *The Routledge Companion to Mobile Media*. Eds, Goggin, G. and Hjorth, L. London: Routledge, 2014, 488–498.

Pope, J. 'A Future for Hypertext Fiction'. *The International Journal of Research into New Media Technologies*, 12 (4), 2006: 447–465.

Prieto, E. *Literature, Geography and the Postmodern Poetics of Place*. London: Palgrave Macmillan, 2013.

Pullinger, K. *Breathe*. Editions at Play, 2018. Web. 5 February 2018. https://editionsatplay.withgoogle.com/#/detail/free-breathe.

Rippner, S. *The Prints of Vija Celmins*. New York: Metropolitan Museum of Art, 2002.

Ritchie, J. 'The Affordances and Constraints of Mobile Locative Narratives'. *The Mobile Story: Narrative Practices with Locative Technologies*. Eds, Farman, J. London: Routledge, 2013, 53–67.

Roberts, L. 'Mapping Cultures: A Spatial Anthropology.' *Mapping Cultures: Place, Practice, Performance*. Ed, Roberts, L. London: Palgrave Macmillan, 2015, 1–25.

Ryan, M. L. *Possible Worlds, Artificial Intelligence and Narrative Theory*. Bloomington, Indianapolis: Indiana University Press, 1991.

Ryan, M. L., Foote, K. and Azaryahu, M. *Narrating Space / Spatializing Narrative: Where Narrative Theory and Geography Meet*. Columbus: Ohio State University Press, 2016.

Sadler, S. *The Situationist City*, Cambridge, MA: MIT Press, 1999.

Sansom, I. 'Whither England? Psychogeography, Walking and Brexit'. *Times Literary Supplement*, 7 July 2017: 32.

Sassen, S. *The Global City: New York, London, Tokyo*. Princeton, NJ: Princeton University Press, 1991.

Saunders, A. *Place and the Scene of Literary Practice*. London: Routledge, 2017.

Sebald, W. G. *Rings of Saturn*, Trans. M. Hulse, London: Harvill, 1998.

Simmel, G. *Simmel on Culture*. London: Sage, 1997.

Sinclair, I. Lud Heat: A Book of Dead Hamlets. London: Albion Village Press, 1975.

Sinclair, I. *Lights Out for the Territory: 9 Excursions in the Secret History of London*. London: Granta, 1997.

Smith, Z. *NW*. London: Hamish Hamilton, 2012.

Smith, Z. 'NW'. *Penguin Books USA*. 2012. Web. 13 August 2018. www.penguin.com/static/pages/features/zadie_smith.

Smyth, K., Power, A. and Martin, R. 'Culturally Mapping Legacies of Collaborative Heritage Projects'. *Valuing Interdisciplinary Collaborative Research: Beyond Impact*. Eds, Keri Facer and Kate Pahl. Bristol: Polity Press, 2017. 191–213.

Soja, E. *Postmetropolis: Critical Studies of Cities and Regions*. Malden: Blackwell, 2000.

Solnit, R. *Wanderlust: A History of Walking*, London: Viking, 1999.

Sorkin, M. (ed.) *Variations on a Theme Park: The New American City and the End of Public Place*. New York: Hill and Wang, 1992.

de Souza e Silva, A. 'From Cyber to Hybrid: Mobile Technologies as Interfaces of Hybrid Spaces'. *Space and Culture* 9 (3), August 2006: 261–278.

de Souza e Silva, A. and Frith, J. *Mobile Interfaces in Public Spaces: Locational Privacy, Control, and Urban Sociability*. New York: Routledge, 2012.

Speakman, D. *It Must Have Been Dark By Then*. Bristol: Taylor Brothers, 2017.

Swatman, R. '*Pokémon Go* Catches Five New World Records'. 10 August 2016. Web. 16 July 2018. www.guinnessworldrecords.com/news/2016/8/pokemon-go-catches-five-world-records-439327.

Tally, R. T. Jr. *Topophrenia: Place, Narrative, and the Spatial Imagination*. Bloomington, Indiana: Indiana University Press, 2019.

Taylor, J. E., Donaldson, C. E., Gregory, I. N. and Butler, J. O. 'Mapping Digitally, Mapping Deep: Exploring Digital Literary Geographies'. *Literary Geographies*, 4 (1) 2018: 10–19.

Townsend, A. M. *Smart Cities: Big Data, Civic Hackers, and the Quest for a New Utopia*. London: W. W. Norton, 2013.

Tuan, Y. *Space and Place: The Perspective of Experience*. Minneapolis: University of Minneapolis Press, 2001.

Turchi, P. *Maps of the Imagination: The Writer as Cartographer*. San Antonio, TX: Trinity University Press, 2004.

United Nations. *Transforming our World: The 2030 Agenda for Sustainable Development*. New York: United Nations, 2015.

United Nations. *2018 Revision of World Urbanization Prospects*. New York: United Nations, 2018.

United Nations Population Fund. *State of World Population 2007: Unleashing the Potential of Urban Growth*. New York: UNFPA, 2007.

United Nations Population Fund. *State of World Population 2011: People and Possibilities in a World of 7 Billion*. New York: UNFPA, 2011.

de Waal, M. *The City as Interface: How New Media are Changing the City*. Rotterdam: nai010 Publishers, 2014.

Wark, M. *The Beach Beneath the Street: The Everyday Life and Glorious Times of the Situationist International*. London: Verso, 2011.

Wilken, R. and Goggin, G. 'Locative Media – Definitions, Histories, Theories'. *Locative Media*. Eds, Wilken, R. and Goggin, G. London: Routledge, 2015, 1–19.

Woolf, V. *Mrs Dalloway*. London: Hogarth Press, 1925.

Zeffiro, A. 'Locative Praxis: Transborder Poetics and Activist Potentials of Experimental Locative Media'. *Locative Media*. Eds, Wilken, R. and Goggin, G. London: Routledge, 2015, 66–80.

Zhang, Y. *The City in Modern Chinese Literature and Film*. Stanford: Stanford University Press, 1996.

7 Collaborative tales

Introduction

I've been using the adjective 'hybrid' throughout this book. At its simplest, the word refers to the condition of being composed of different elements. Yet this state of existing across more than one discrete domain or entity has been key to my definition of contemporary existence. Postdigitality recognises that the binary divide between digital and non-digital domains no longer holds and that in its place has come a fundamental entanglement of digital and non-digital, physical and subjective, embodied and textual. The creative works discussed throughout this book have shown how writers and artists have responded to this evolving condition, exploring innovative ways of both representing and interrogating, what amounts to, a new post-postmodern ontology.

These ideas and concepts naturally coalesce around the most iconic object of the early twenty-first-century: the smartphone. As we've seen, writers have seized on the opportunities afforded by locative mobile technology, developing app-based narratives that have begun to investigate the potential of new forms of embodied, situated storytelling. App-based works such as James Atlee's, *The Cartographer's Confession*, or Duncan Speakman's, *It Must Have Been Dark By Then*, are part of a growing praxis into the nature and form of such storytelling. What I have termed *embodied space* is just one aspect of this, of course, a new hybridic form of narrative space created through the interplay between the digital and the non-digital, the embodied and the textual.

What I want to do in this chapter is develop and extend this argument. While not sidelining the sort of locative mobile storytelling we've already examined in Chapter 6, I want to purposely build on these foundations. Specifically I want to discuss the newfound opportunities for collaborative or collective storytelling that digital technology has opened up. Collaborative, community-based practice has a long and established history within digital storytelling, of course. In *Digital Storytelling: Capturing Lives, Creating Community* (2013), Joe Lambert notes the central role that storytelling plays in how we engage with, and understand, the world around us:

> As suggested, story has many jobs, as a learning modality through memory, as a way to address our connection to the changing world around us, as a form of reflection against the flood of ubiquitous access to infinite information, as the vehicle to encourage our social agency, and finally, as a process by which we best make sense of our lives and our identity.
>
> (2013, 14)

Lambert's emphasis on the affective, situated and embodied nature of storytelling fits neatly with the argument I've been developing across this book. Stories are a way of 'worlding the world', as Heidegger would have it, a form of wayfaring through Tim Ingold's meshwork of interwoven trails (2016, 93). But as organisations such as *The Center for Digital Storytelling* demonstrate, these stories do not necessarily need to be individual or discrete in their conception or intent. For Lambert, it is the participatory, community-based potential of digital storytelling that is of particular value. Yet, as Mark Dunford and Trisha Jenkins also note, the challenge remains of understanding how such approaches can fully capitalise on ongoing technological developments while at the same time remaining true to the value of community-based, participatory storytelling (2017, 14).

Such developments, of course, include the wide range of platforms that come under the umbrella term, social media. Twitter, Instagram and Facebook, as well as Web 2.0 functionality such as blogs and wikis, have transformed how we engage with each other. According to the Office of Communications (Ofcom 2018), in 2018 more than three-quarters of UK internet users (77%) now use social media, which equates to two-thirds (68%) of adults overall. Globally, the data is even more striking, with an estimated 3.5 billion active social media users across the planet (Kemp 2018), representing a growth of thirteen per cent on the previous year (Ofcom 2018), with much of that in the developing world. The sort of storytelling that takes place on these platforms is complex and wide ranging. Ruth E. Page is surely right to state that the growth of social media has allowed people to 'document the stories of their daily experiences in online, public, or semi-public domains in unprecedented measure' (2012, xv). Understanding this new form of storytelling, though very much in its early stages, remains one of the most important challenges for any study of contemporary storytelling. The US 2016 presidential election and the UK 2016 EU referendum are just two recent events in which the role of social media has come to be seen as controversial. Never before have the stories we tell each other, and the means by which we tell them, mattered so deeply. Yet, although undoubtedly mired in controversy, it is the contention of this chapter that social media can make a positive and meaningful contribution to our lives. Their spread and penetration is just too significant for academic researchers to either ignore or sideline. As we'll see, beneath the headlines, real and affective progress is being made in terms of how social media platforms can be used to craft new forms of narratives, stories that are 'emergent, collaborative, and context-rich' (Page 2012, 12). A good example of the sort of thing I mean here is the #ohneMauerfall Twitter campaign. The 6th February 2018 marked the 10,316th day since the fall of the Berlin Wall. It was also the point at which Germany's postwall period had lasted longer than the time the wall had been in place. Ohne Mauerfall is German for 'without the fall of the wall'; on the 6th February Germans used the hashtag #ohneMauerfall to share their reflections and thoughts on Twitter about how different their lives would have been if the wall had not come down when it did in November 1989. Thousands of individuals from across the globe took up the challenge, all threaded together into a single narrative through the hashtag.

One should be cautious of drawing too many conclusions from a discrete example such as this. It should also be stressed that the concept of 'writing on the city' is nothing new: graffiti, written by a literary elite and inscribed on statues and government buildings, was not unusual in ancient Rome. Still, in many ways, the #ohneMauerfall campaign offers a useful insight into the potential value of social media. #ohneMauerfall was completely democratic and open, the only restrictions imposed by those of the platform itself. The campaign spread rapidly across the Twittersphere, requiring nothing else than the vast network of interlinked users. Its success in that sense was down entirely to personalised action and response. The hashtag turned the personal thoughts expressed in each individual tweet into a collaborative, dialogic narrative that was both open and emergent in real time. Retweets, likes and comments, as well as links and embedding, enabled even deeper levels of interactivity and enmeshment. Photographs and geotagging provided a spatial specificity to many of the tweets, a way of visually celebrating the freedom to occupy physical locations that would have been excluded. The end result of #ohneMauerfall was a powerful example of participatory, and affective, storytelling that hovered between fact and fiction where the online identity of participants was deeply amorphous. I would argue that both the act of writing and publishing a tweet, and reading what others had written, were a means of knowledge (co)-creation, an invitation to 'think differently', as Facer and Pahl describe it (2017, 226).

These characteristics, what Page describes as collaborative, dialogic, emergent and context-rich (2012, 8), will underpin much of what I have to say in this chapter. The collaborative affordances of social media, in particular, have provided an environment in which new types and forms of individual and collective storytelling is emerging. Campaigns such as #ohneMauerfall indicate the way artists and writers are utilising such functionality to create innovative opportunities for affective narrative interventions. Lyle Skains uses the term demotic author to describe this phenomenon, where the author has become '"one of the people," participating in a community of writers and readers' (2019, 2). What is clear already, however, is that understanding such work from a purely digital perspective is limiting and unhelpful. #ohneMauerfall was made possible by a digital platform, of course; yet to label it a digital story is misguided. The fact that a significant percentage of the tweets were written on a smartphone, out on the street, on location, suggests that #ohneMauerfall shares a lot of similarity to the locative mobile stories we examined in the previous chapter. If this is the case, then the sorts of storytelling we see on social media platforms are more usefully described as postdigital. The #ohneMauerfall tweets were not simply the product of digitality; the physical location of the writer often played an instrumental part in the composition of the tweet itself; indeed, the mobile affordance of the smartphone actively encouraged a kind of situated embodiment, in which the physical location of both writer and reader became an overt element of the participatory experience. This fundamental postdigitality of social media, what Page recognises as its essential hybridity and intertextuality (2012, 15), offers new and innovative ways of thinking about these kinds of stories. And as writers such as Zadie Smith, Jon McGregor and Joanna Walsh demonstrate, this influence is not limited to the digital domain but reaches out into offline, print-based works too.

Such hybridity has important implications for arts and humanities research. Whereas Facer and Pahl are right to extoll the value of interdisciplinary and collaborative research in projects such as the Cultural Mapping Toolkit (Smyth *et al.* 2017), it is also clear that outward facing, collaborative, community-situated engagement with both academic and non-academic stakeholders, requires new approaches too, allowing the exploration of what Jay Bernstein calls 'the inherent complexity of reality' (2015, 13). This concept of transdisciplinarity, with its emphasis on Heidegger's notion of 'being-in-the-world' or what Alfonso Montuori terms 'in vivo' (2008, xi), brings both this chapter, and the substantive part of the book, to a close.

Digital storytelling: from one to many

The term digital storytelling is one that I've been treating with a degree of caution throughout this book. From the perspective of postdigitality, digital story has an unsatisfactory simplicity, disguising as much as clarifying the inherent complexity behind how these stories are written and experienced. If one rejects the essential binarity of the digital/non-digital domains, then a term such as digital storytelling becomes equally problematic.

Yet there are areas where the term has become indelibly associated with a certain type of storifying. The non-profit organisation, the Centre for Digital Storytelling (CDS), is a good example of this. According to its founder, Joe Lambert, at the heart of the CDS lies the democratising power of storytelling. Here, storytelling takes on a therapeutic role, a means by which individuals can come to terms with emotionally complex and difficult events. As we've seen, Lambert uses the term 'learning modality' (2013, 14) to describe the role that storifying plays in this regard: in other words, making stories can be understood as a means through which we gain understanding and insight about ourselves and the world around us. Yet it is also an overt exploration of ambiguity, the co-construction of meaning. As Lambert explains:

> What story cannot do is completely simplify the messiness of living. Story is essentially an exercise in controlled ambiguity. And given the co-constructed nature of meaning between us as storytellers, and those who are willing to listen to our words, this is the story's greatest gift.
>
> (2013, 14)

At the CDS, these kinds of stories break down into a variety of tropes, including stories about people (what Lambert calls character and memorial stories), events, place and activity (2013, 21). Around these tropes hovers other forms, such as recovery, discovery and coming of age narratives.

None of this necessarily dictates the use of digital media. Creative writing as a therapeutic intervention has a respected history, much of it undertaken with traditional pen and paper. Yet, for Lambert, digital media offers unprecedented opportunity for ordinary people in terms of constructing their own stories, and getting those stories read by other people. The key is establishing a structure that guides participants, some of whom may have limited technical skills, safely through the production process. For the CDS, the physical, face-to-face workshop remains paramount. Lambert recommends a three-day process by the end of which participants have designed, constructed and shared their digital work (2013, 75). At the heart of this process is what he calls a story circle (77). During the story circle, participants receive feedback on their preliminary idea for a story. What each participant then produces as a digital story is very much informed by this collaborative process. Indeed, for both Brooke Hessler and Joe Lambert, this collaborative imperative is fundamental to how they understand and use digital storytelling within the CDS; as they state, '[d]igital storytelling was conceived as a group dynamic informed by skillful facilitation' (2017, 26). This 'group dynamic', as they put it, is built around the notion of the story circle.

Hessler and Lambert use Jan Meyer and Ray Land's notion of threshold concepts to explore the radical potential of digital storytelling as means of providing new ways of seeing and knowledge creation. A threshold concept is similar to a portal, opening up a new and previously inaccessible way of thinking about something. As Meyer and Land go on to explain, a 'new way of understanding, interpreting, or viewing something may thus emerge – a transformed internal view of subject matter, subject landscape, or even world view' (2005, 373). They go on to explain that these conceptual portals hold a number of key characteristics: they are transformative in that they engender a significant shift in the perception of an individual; their effect is irreversible in that it is unlikely to be forgotten or ignored; it is integrative, in that it reveals the previously hidden interrelatedness of something; and, finally, it may lead to, what Meyer and Land call, 'troublesome' or difficult knowledge (2005, 373–374).

Hessler and Lambert argue that their collaborative model of digital storytelling is a fundamentally transformative, dynamic and radical process that explicitly invites threshold crossing. During the CDS process, participants are encouraged to understand the 'fundamental authority on their own personal experience' (2017, 32). The starting point is always the situated and embodied experience of the individual, and it is from the recognition and understanding of this personal and individualised perspective that the digital story develops.

The result of all this is that the primary focus of the CDS' approach is very much focussed on the benefits accrued by participants. The hosting and the subsequent reading of the works by others remains very much a secondary concern. Indeed, as Lambert explains, the final showcasing of any digital story is very much left to each participant (2013, 119). Instead, the emphasis is on the experience of the writer in which the construction of a digital storytelling operates as a form of self reflection. This approach, including the story circle, has been brought together into seven clearly definable steps. Taken together, they form the methodological spine of the CDS' approach. Lambert explains that the Seven Steps are intended to help participants visualise their intended story in terms of architecture and media (54–69). Step 1 is called 'owning your insights' (54). Self-reflection allows each participant to question the real meaning of their story and then, through this insight, to redefine their original ideas. Step 2 (owning your emotions) encourages participants to fully recognise their own emotional connection to the story they're telling; step 3 (finding the moment) encourages participants to locate the dramatic focus of their story, the key scenes that best and most succinctly represent what they are trying to say; step 4 (seeing your story) and step 5 (hearing your story) invites participants to pinpoint the central imagery and sound to be used, such as photographs and voice-over; step 6 (assembling your story) brings into play questions of structure through scripting and storyboarding; and finally step 7 is the possible sharing of the story, on the CDS website, for example, or through a live presentation.

The digital story 'Participant-Observation' by Wynne Maggi, hosted on the CDS website, is a good example of the sort of story participants have produced.[1] The story, lasting four and a half minutes, is told through the voice over of the author, accompanied by a series of photographs. The intimacy of the story, recalling the death of a baby boy while the author was undertaking field work in Pakistan, is subtly enhanced by the voice, the aesthetics of the photographs and the minimalist register of the script. The overall effect is one of an intense search for meaning and understanding, a personal meditation on something that still resonates across the author's life. This is a poetic effect, something short story writer's in particular would be very familiar with. As Cynthia Hallett notes:

> As a rule, writers of the minimalist short story manipulate figurative speech to present what appears to be a single event, a mere incident, a

> nothing-is-happening-here story that is actually an intricate figurative pattern that reflects or signifies the human condition and capacity.
>
> (1999, 16)

The retrospective narration of 'Participant-Observation' deliberately avoids directly addressing the effect that the death of the boy has had on the author. Instead, it is pushed into the subtext of the story, powerfully evoked through the very last line in which we're simply told, 'I never went back to the field'. This is classic short storytelling. Yet perhaps we shouldn't be too surprised: elements of the Seven Step process could come straight out of a writer's guide to modern short fiction, particularly the emphasis on subtext and emotional meaning. 'Participant-Observation' is undoubtedly a good example of the way digital stories can help participants come to terms with 'troublesome' knowledge. According to Lambert, Wynne Maggi broke down in the original story circle after relating the incident, exclaiming, 'I've never told anyone that story before' (2013, 52). Yet, removed from the context of the CDS website, the story could be seen as a far-more ambiguous work, in which its autobiographical underpinnings are less than certain. None of the photographs in 'Participant-Observation' are labelled, even those that purportedly show the author; the identity of the spoken-voice is equally left unidentified. The very title itself hints at the dangers of assumption and interpretation. These complexities lurk on the edges of the narrative, even when we know the context of its making. Clearly, they are also calculated authorial decisions that have emerged through the Seven Step process. In other words, it's not necessarily the autobiographical 'realism' of 'Participant-Observation' that generates its affectivity but rather the power of a well-told story.

This kind of digital storytelling is not limited to the CDS, of course. As Dunford and Jenkins demonstrate, examples can be found across the world. Yet, as they also note, any optimism about its popularity is also inflected with a certain frustration: collaborative projects tend to be small and insular in which any impact 'either individually or collectively remains arguably limited' (2017, 14). There are exceptions, of course. The *One Million Youth Life Stories* initiative, which ran between 2006 and 2008 in Brazil, was an attempt to draw an entire demographic into the political process of the country (Worcman and Garde-Hansen 2016, 120). Young people in Brazil have particularly high associations with social vulnerability, unemployment and violence, and remain disaffected and poorly represented in political forums. The *One Million Youth Life Stories* initiative was set by a number of social organisations with funding from the Kellogg Foundation. As Worcman and Garde-Hansen explain, the project was predicated on the concept that encouraging young people to share their stories would be an effective way to encourage social and political participation and engagement (120). The project explicitly drew on the methodology created by the CDS; in the two years that the project ran, *One Million Youth Life Stories* facilitated ten workshops with three hundred participants; these participants in turn ran a further one hundred workshops, involving three thousand people (Misorelli 2017, 38). Just like 'Participant-Observation', the stories were based on the use of image and voice-over. A key aspect of the project was the dissemination of the stories. Participants were encouraged to publicise their stories themselves through social media and local screenings; a DVD was also produced in which the issues raised by the participants, including drug and alcohol abuse, were brought together. Some of these were then sent to the deputies in the National Congress of Brazil. As Misorelli recognises, through the project 'young people were able to discuss their social context, youth public policies and community development' (2017, 39).

The CDS model of digital storytelling is clearly an important and influential one. The Seven Step process, with its emphasis on the story circle, is empowering and supportive. The methodology places co-creation and mutuality at its heart; stories emerge through open discussion and debate. The end result is not always what the participant had originally intended. While work such as 'Participant-Observation' are singular and individual in their construction, initiatives such as *One Million Youth Life Stories* indicate how the approach can also be used across communities and groups to build social and cultural cohesion. Yet, in both cases, the primary emphasis remains on the collaborative *construction* of the digital stories rather than their *dissemination* as texts. Indeed, as we've seen, for Lambert, the sharing of the stories is the most difficult and sensitive aspect of the entire process. This construction draws on established technologies to produce a web-based multimedia presentation that relies heavily on image, text and audio. This is not to denigrate these digital stories but rather to note that, in their own way, they reflect the methodology's emphasis on *process* as much as *output*. From personal experience, through to the co-creative environment of the story circle, the CDS model of storytelling is inherently hybridic, a celebration of the collaborative and the embodied. As Lambert states, the CDS approach is not about becoming an author *per se*; rather, it explores how authorship can be used to create affective agency and life-affirming social interaction (2013, 2). In the journey from the story circle to the digital output, the divide between the digital and the non-digital is collapsed. What we are left with, then, is a mode of storytelling that is inherently postdigital in its philosophy and approach.

Social media

While the CDS remains very much wedded to the production of standalone multimedia stories, more recent technological developments have begun to open up alternative approaches. Some of these we've already looked at, including locative mobile storytelling and contemporary hypertext stories (such as Joanna Walsh's *Seed*, 2017, for example). Another related area that I'd like to explore here is the impact of social media platforms on storytelling.

Social media is a rather nebulous term. However, in essence it is a noun that refers to the range of applications that support content co-creation and sharing, and social networking. These applications include blogs, wikis, podcasts, discussion forums and social networking sites. This real-time co-creativity of content, as well as the ability to embed and share that content, offers storytelling a radical new set of affordances that clearly demarcates it from the CDS model. Perhaps the biggest difference, however, is the fact that social media have become ubiquitous, 'an integral part of everyday life' (Hinton and Hjorth 2013, 2) where the reach and dissemination of social media content is vastly expanded. Crucially, as we've seen elsewhere in this book, this influence also extends not only *within* the dominant digital domain, but also *across* into the subordinate domain of print-based forms such as novels and short stories (see Chapter 3 for more detail on this).

For some, this ubiquity only reinforces the opinion that social media should be viewed with deep cynicism. In his book chapter, 'Social Network Exploitation', Mark Andrejevic cautions against euphoric optimism, arguing that the use of user-generated content and data by corporate technology companies amounts to a new form of capitalist exploitation.

> Contrary to conventional wisdom, social networking sites don't publicize community, they *privatize* it. Commercial social networking sites are ostensibly collaborative productions, except when it comes to structuring terms-of-use agreements, and, of course, allocating the profits they generate.
>
> (2011, 97, italics in the original)

These concerns have only escalated as companies such as Facebook have become embroiled in controversy surrounding the corporate use of personal data.[2] Moreover, the US 2016 presidential election and the UK 2016 EU referendum have illuminated some of the weaknesses in the way social media works, particularly the ghettoisation of news streams through homophilous sorting where like-minded people form enclosed 'ecosystems', resistant to alternative points of view and opinion (Ancona 2017, 50).

For others, however, the situation is more equivocal. Henry Jenkins, for example, argues that social media is just part of a wider 'convergence culture' (2006, 2–3). By this he refers to the social and cultural effects brought about by the increasing interdependence of technological systems, processes and content. It is this very convergence that encourages the growth of an empowering participatory culture and, what Jenkins terms, 'collective intelligence' (2006, 245). In a similar vein, Sam Hinton and Larissa Hjorth note the way social media embraces new forms of social intimacy through the sharing of personal information, leading to the creation of, what they term, 'intimate publics' (2013, 44).

In their book, *The Ambivalent Internet* (2017), Whitney Phillips and Ryan M. Milner posit a middle way through these two oppositional positions. Their argument rests on two assumptions. The first is that online expression is neither good nor bad, enlightening or reactionary. Rather, such expression is ambivalent, with the emphasis on the Latinate prefix *ambi* 'which means "both, on both sides"' (2017, 10). In other words, online materials 'are not singular: they inhabit, instead, a full spectrum of purposes – all depending on who is participating' (2017, 10). Their second argument is that these online behaviours can only be properly understood as extensions of offline culture; as they go on to state, '[o]nline and offline experiences are in fact so fundamentally intertwined that it's impossible to parse where the embodied ends and the digital begins; the one sustains and contextualises the other' (2017, 66–67). Hinton and Hjorth concur on this point, acknowledging that social media is never just about a digital phenomenon but rather a complex set of online/offline entanglements: '[t]he relationships that people have online are always shaping, and shaped by, the offline (2013, 3).

In many ways, we are all still finding our way through these debates. It is certainly true that the rise of social media has been controversial, particularly through its association with the rise of political extremism and voter manipulation. Yet, concurrently, research also shows that there are real benefits to be had. At the very least, the concept of an ambivalent internet, in which the onus of responsibility is pushed back on us, seems appealing. Yet the formation of 'intimate publics' through online interaction, such as the #ohneMauerfall Twitter campaign, points to a more purposeful interpretation. It is here that this chapter aligns itself, embracing the view that digital technology has real and tangible benefits for artists and writers. In particular, the value of social media as a way of supporting and sustaining collaborative, community-based interventions will be highlighted, aligning much of what I have to say alongside Facer and Pahl's notion of embodied learning and productive divergence (2017, 17). Critical here will be the way social media can be understood from the perspective of the postdigital: as inherently entangled, embodied and hybridic.

Social media as story

The development and growth of social media is the most significant outcome of Web 2.0. We've already looked at the Web 2.0 phenomenon in Chapter 4, so a brief summary is only needed here. Web 2.0 emerged in the early years of this century; whereas early versions of the web tended to be static and uneditable, Web 2.0 offered enhanced user participation through dynamic content editing and sharing. It was this development that provided the technical foundation on which social media platforms were subsequently built. Facebook was launched in 2004, Twitter in 2006 and Instagram in 2010. These sites offered users the ability to add, edit and share multimedia content on the web in ways that were previously impossible. With the release of the first iPhone in 2007, and the subsequent development of mobile apps from 2008, social media took on a whole new level of functionality and user engagement. It was from this moment onwards that platforms such as Facebook, Twitter and Instagram became an indispensable part of life, never being any further away than the phone in our pocket. If *We Are Social/Hootsuite's* data is correct (Kemp 2018) and there are indeed 3.5 billion active social media users, then it means that almost half the world's population is, in some way or other, a user of these platforms. This is quite astounding given that Web 2.0 did not begin to emerge until 2002.

Page usefully reminds us that social media did not spring up in a vacuum. Online discussion boards and forums[3], short message service (SMS) and even email can be seen as early precursors of what was to come. Yet despite these continuities, it's important not to downplay the sense of disjuncture. In terms of affordance and scale alone, the continued rise of social media is a social and cultural phenomenon unprecedented in human history. Page lists five features that make social media a distinct and separate experience: social media environments are collaborative, dialogic, emergent, personalised and context-rich (2012, 8). They are collaborative in the way they reach out to a potentially limitless audience; they are dialogic in that they are constructed around collaborative conversation or dialogue, in which a lot of content is a response to other content; content is emergent, generated in real time across multiple platforms and audiences; such content is also personalised and personalisable, both in terms of authorship but also how it appears; and finally, these characteristics give rise to an environment in which content is fundamentally recontextualised, shared and embedded across multiple platforms and devices with no dependency on time and geography.

In terms of storytelling, these five characteristics have some important consequences for authors. As Page notes, '[t]he emergent stories of social media are often open-ended, discontinuous, and fluctuating' (2012, 12). This open-endedness is in contrast to the episodic form of social media stories, engendered by the meshwork of posts, tweets, replies and updates. By default, these are often published in reverse chronological order, with the newest posts at the top, foregrounding immediacy and a kind of narrative, or authorial, 'presentism'. #ohneMauerfall offers a good example of this: the tweets were emergent in real time, with the newest at the very top. This dialogic montage of thought and memory, text, link and image, was then supplemented through synchronous likes, retweets and replies leading to the formation of Hinton and Hjorth's intimate publics (2013, 44).

Yet the emergent and collaborative nature of social media stories also means that the wider context in which those stories are received is fundamentally dynamic. Authors are used to the idea that they have no control over where or how a traditionally published text, such as a novel or poetry collection, is read. However, the very form of a hardcopy work ensures the actual content cannot be altered; in most cases the design, structure and ordering is set. Yet these boundaries do not apply to the content published through social media. The shareability and embeddedness of any content on these platforms introduces a unique set of contextual issues for any authors of social media. Perhaps the most obvious is what Page calls the *textual context* of social media (2012, 15), in other words, the dynamic instantiation of content from numerous sources. Authors using #ohneMauerfall, for example, could not determine the ordering or position of their tweet within the live feed. Nor could they control any retweets, comments or likes of their work. Even the interface itself was not prescriptive: individual tweets or even the entire #ohneMauerfall feed might be embedded within another social media platform or web page with its own screen layout, interface and underlying discourse. For #ohneMauerfall this might not be a particular concern; for a more overtly political campaign such as #metoo, however, the way tweets, posts and blogs are shared and embedded potentially becomes much more of an issue.

Page notes two other contextual issues for social media that are relevant here (2012, 15). The first is what she terms the extrasituational context. By this Page refers to the social-cultural context of the participant, including age, gender and ethnicity. The second, behavioural context, is the actual physical location of the participant. Not only are both of these interconnected, they also open up the possibility of what I have termed 'embodied space' as a narrative trope, in other words the explicit interplay between the physical and textual worlds. I'll be exploring this later on in the chapter. For now it's worth saying that these extrasituational and behavioural contexts introduce a dynamic fluidity within social media that suggests the notion of an 'intimate publics' needs careful thought. As we'll see, the technology offers a degree of shareability and embeddedness that means no one has ultimate control over where or how it is used. Instead it is perhaps better to consider social media as offering the potential for countless, coterminous publics, some of which might be labelled intimate, with others operating more transgressively or oppositionally. As Page notes:

> While the emergent nature of social media promotes episodic narrative sequences, the individual units of social media stories (posts, updates, or tweets) do not have to be interpreted atomistically, but are positioned within broader generic, discourse, and behavioural contexts that are underpinned by networked connections between narrators and their audiences.
>
> (2012, 196)

For Phillips and Milner, this 'ambivalence' or diversity of voice across social media is to be welcomed and in many ways reflects offline or real world behaviour. Yet, ultimately, such extrasituational and behavioural contexts merely reinforce what has already been said in this book: postdigital storytelling, including that undertaken through social media, can only be properly understood as occurring through an entanglement of offline and online domains. 'Storytelling may take place via screens, but is produced, received, and interpreted in contexts that bridge online and offline environments' (Page 2012, 205). In other words, the concept of the postdigital developed across this book remains at the heart of how we should understand and critically explore social media, not only in terms of storytelling, but also in its wider societal impact.

Twitter as story

Twitter provides an excellent case study for these issues. Unlike Facebook, which has always been primarily an informal, even convivial, space, Twitter offers a much more functional, even didactic, environment. Founded in 2006 by Jack Dorsey, Twitter positioned itself as a new kind of microblogging site, in which each blog, or tweet, was limited to 140 characters or less. As well as text, hypertext and images, tweets can also include hashtags. Hashtags are a way of categorising tweets so that others can find them. Tapping or clicking a hashtag within a tweet will bring up all the other tweets that have also used that hashtag. A hashtag can be completely functional, such as #postdigital, or it can be an essential part of the tweets intended meaning, such as #metoo. In the latter case, the hashtag does not simply categorise; in many instances, it fundamentally transforms how the tweet is interpreted.

Three more recent developments have enhanced Twitter's storytelling functionality: the doubling of the character restriction, the ability to create threads and the ability to create 'moments'. All three developments indicate that the drive to storify on social media is not simply driven by user experimentation. In the case of Twitter, the platform itself has noticeably shifted, embracing new functionality that makes storifying easier. As well as an enhanced 280 character tweet, threading allows users to link or connect a series of tweets. A user can bring tweets together and then publish the thread in its entirety using the 'tweet all' option; alternatively a tweet can be added to a thread by the user at any time, allowing the thread to grow over time. In comparison, a moment is a curated story, brought together by stitching together existing tweets. A moment has a different look and feel to the traditional Twitter interface, existing more as a kind of multimedia story with its own separate page and aesthetic.

The qualities that we've already identified, namely collaborative, dialogic, emergent, personalised and context-rich (Page 2012, 8), are still very much in evidence in this new functionality. What it does, however, is present authors with a more conventional range of storytelling options. In July 2014 the author David Mitchell began tweeting his short story, 'The Right Sort'. Consisting of 280 individual tweets sent out over seven days, the story is narrated in the present tense by a boy hallucinating on his mother's Valium pills. Predating threading, the story is structured in reverse order, with the very end of the story at the top of the sequence, and the beginning at the bottom. To get to the beginning, one has to scroll backwards. For those reading the story live this wasn't an issue. Yet for anyone coming to the story late, or wanting to read it today, 'The Right Sort' has a curious tension between the structural architecture imposed by the medium (in reverse), and narrative architecture of the story (forwards). The end of one is the beginning of the other and vice versa, reinforced by the time and date stamps on each tweet. In an interview with *The Guardian* newspaper, Mitchell noted that, what he termed, 'narrative tweets' needed to be like haiku, in the sense of having their own inner logic or 'rightness', while also having narrative momentum (Flood 2014). Yet he might also have noted that the very format of his story was a record of its own 'liveness': scrolling down through the tweets to the beginning is also a movement through the temporal (real-time) evolution of the story.

Jennifer Egan's 'Black Box' is another famous example.[4] Egan's story was published over nine days in May 2012 and tells the story of a female spy in the near future who has covert recording apparatus implanted into her body. The black box of the title is the female spy, and Egan's work, at least in part, explores how women are objectified through technology. Rather like Mitchell's 'The Right Sort', Egan's story is exploratory, even experimental, in its use of Twitter. It's written in second person using future simple and future perfect continuous. In that sense the story operates through a kind of personal address to the reader: 'When you know that a person is violent and ruthless, you will see violent ruthlessness in such basic things as his swim stroke' (Egan 2012).

Both 'The Right Sort' and 'Black Box' embrace the emergent, dialogic and personalised aspects of social media. These aspects are noticeably heightened by the form of the writing itself, particularly perspective and tense. Yet the stories are also collaborative and context-rich. Readers of both stories actively engaged in the storytelling process itself through likes, replies and retweets. The extant story is as much a record of these collaborations and interactions as it is of the formal text written by the author.

The writer, photographer and art historian, Teju Cole, has taken a different approach in his use of Twitter through his small fates project. In 2011 he began tweeting a series of seemingly obscure stories about ordinary people living in Nigeria. For example, on the 19 August 2011, Cole tweeted the following: 'Mrs Ojo, of Akure, favoring a cashless economy, bought goats, beans, and onions in Sokoto, and paid with bags of marijuana'. It was one of many such tweets, all seemingly random and unrelated to each other: ' "Nobody shot anybody," the Abuja police spokesman confirmed, after the driver Stephen, 35, shot by Abuja police, almost died' (29 June 2011). Cole explained that he had adapted the text of each tweet from Nigerian newspapers where columns of such stories – brief, factual and elliptical – would appear as a kind of smorgasbord of everyday drama (Cole 2011). Cole notes how, in French, these sorts of factual sundry events are termed faits-divers. Their fragmented, seemingly meaningless nature, told in deadpan reportage, surfaces the minutiae and the ordinary in a globalised world. They offer a kind of challenge to the reader, especially when brought together en masse, forcing them to re-engage with notions of story, meaning and affect. Cole is not the first to have recognised the affective power of faits-divers. In 1906 the French writer and anarchist, Félix Fénéon, wrote thousands of anonymous three-line news items for *Le Matin*. Saved in Fénéon's own scrapbook, they were only finally published in 2007, under the title, *Novels in Three Lines*. As Luc Sante notes in his introduction, '[i]f each item is a miniature clockwork of language and event, the full thousand-and-some put together make a mosaic panorama' (2007, xxvii), conjuring up the year 1906 in a way that no official history has ever done. Cole's small fates project brings *Novels in Three Lines* into the twenty-first century. Cole adapted the faits-divers he found in the Nigerian press, stripping them down and re-editing them so that they would work within the 140-character constraint of Twitter. As Cole explains:

> Each tells a truth, a whole truth, but never the whole truth (but this is true of all storytelling). Details are suppressed, secondary characters vanish, sometimes the 'important' aspect of the story is sidestepped in order to highlight a poignant detail.
>
> (2011)

Cole argues that these stories offer insight into the postcolonial modernity of Nigeria. Yet they also provide a mirror by which to see the micro-stories of social media, connecting with us and distancing us in equal measure. The small fates reach out across the internet offering a challenge to affectivity and truth-making, just as Fénéon was doing in the early years of the twentieth century: 'Eugène Périchot, of Pailles, near Saint-Maixent, entertained at his home Mme Lemartrier. Eugène Dupuis came to fetch her. They killed him. Love' (Fénéon 2007, 40).

Taken together, Cole's small fates project, alongside Twitter fiction such as 'The Right Sort' and 'Black Box', provide a useful insight into the narrative possibilities of social media platforms, such as Twitter. Although more recent functionality such as threading and moments has augmented the storifying capabilities of Twitter, the three examples given here show the degree to which the simple unreconstructed tweet can become the bedrock of narrative form. Without hashtags, web links or images, the tweets maintain a strong affinity with the printed page. This fragmented approach to narrative is something we've seen in Ruth Tringham's *Dead Women Do Tell Tales* project, discussed in Chapter 1. The digital platform allowed the creation of, what Tringham calls, 'recombinant histories', made up of narratives that were 'collections of indefinitely retrievable fragments' (Anderson 2011). Although not developed for social media – there are no intimate publics here, as we've defined the term – Tringham's project nevertheless embraces a form of historical narrative that is deliberately emergent, dialogic and personalised, foregrounding interpretative subjectivity and partiality.

However, other possibilities remain. In particular I'd like to explore what Page has called the behavioural context of social media and what I have termed embodied space. In other words, I'd like to push beyond thinking about a tweet as simply text and instead engage with some of the ideas I introduced in the previous chapter regarding locative mobile storytelling. To what extent does social media offer new ways of thinking about embodied space as a narrative phenomenon and what impact do such developments have on our understanding of postdigital storytelling?

Walkways and Waterways

To help answer these questions I'd like to analyse a project, *Walkways and Waterways*, that specifically explored the real-time interaction between user-generated content and urban space.[5] The project was a collaboration involving Cardiff Metropolitan University, the University of South Wales and the digital media startup, *Fresh Content Creation*. The purpose of the project was to explore how creative exploration, mediated through mobile technology, could help users re-explore their own cityspace. An earlier unrelated project, 'Walking Through Time', had showed what could be done in terms of the real-time tracking of individuals across the virtual representation of historical maps on smartphones. As Chris Speed states, 'being "in" the map and being located on a street that no longer exists offers new methods of understanding our surroundings' (2015, 177). Another project, *I Tweet Dead People* (funded through Research and Enterprise in Arts and Creative Technology, UK) had successfully proved that Twitter could be used as a means of guiding people through a particular space. In this case, a digital trail was set up in York Museum, UK, which both guided visitors along a series of 'clues' as well as being used to activate multimedia installations.

Walkways and Waterways wanted to extend these approaches by using Twitter to capture the creative responses of participants as they physically progressed across the city. In that sense, *Walkways and Waterways* was much more focused on the relationship between narrative and physical movement, in other words, the creative and affective potential of what I have termed embodied space.

The project itself was based around a single event consisting of a digitally-mediated journey that retraced the last two miles of the Glamorganshire Canal from Cardiff Castle to Cardiff Bay (Figure 7.1). The Canal itself fell into ruin soon after the Second World War and has long since been removed. Retracing it through the centre of a large city involved traversing a variety of terrains, including a modern shopping centre, before finishing at the location of the canal's old sea lock, beneath a busy flyover.

Participation in the single-day event was advertised through social media. Twenty participants attended, ranging from young children to the retired. While the physical journey was led by project members, exploration was also digitally augmented by Twitter through which further guidance and information was given to each participant in real time. The smartphone allowed each participant to upload photographs and commentary to #GlamCan, providing a shareable forum through which every individual itinerary could be recorded and collated. Crucially, participants were also encouraged to activate Twitter's geotagging feature so that each tweet would include locationary data. This metadata also acts rather like a hashtag in that, by clicking on it,

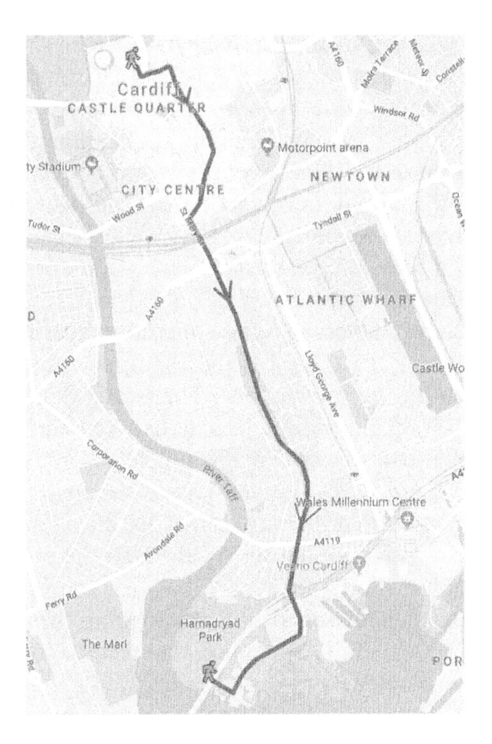

Figure 7.1 Walkways and Waterways: the route of the walk.
Source: The author, map data ©2019 Google.

a user can bring up a list of other tweets that have also been tagged with that geolocationary position.

Embedded within the walk were twelve 'treasures', forming, what was described as, a 'treasure trail'. In this way, *Waterways and Walkways* experimented with real-time gaming and play as a way of enhancing participation and engagement. Tweets sent by the project team prompted participants to both find and then record the next treasure. Sometimes, as in the case of a marooned paddle post in the subway beneath a dual carriageway, participants were invited to discuss its function. Participants then uploaded their responses to Twitter. Sometimes team members responded to particular replies; participants also engaged in online discussion with each other.

Specific directions sent through Twitter guided the participants across the city: 'Walk through the old tunnel, follow the guides. Turn to your right at Hilton and look up/Go into the arcade opposite, take stairs on the left, walk down & follow guides; this is tricky water to navigate!' Old photographs were tweeted to the participants at key locations, asking them to compare the existing topography with what was recorded in the photograph. Participants were encouraged to respond through their own tweets which were then shared across the group. Sometimes these responses were simple factual responses to the questions: 'old tunnel'; 'custom house unchanged'; 'rope'. At other times a more abstract, creative response was forthcoming that captured the participant's fleeting emotional responses: 'Dark, cold, wet. Voices echo in tunnel/swimming along the canal through tides of shoppers'. And just occasionally, memories suddenly surfaced: 'Remember canal as a child. Memories of swimming in it/Met my husband here – 40 years ago!'

In other situations the limitations of a tweet could be considered unnecessarily restrictive. Yet, out on the street, the brevity and concision imposed by the medium became a real strength. Each tweet offered the opportunity to add up to four photographs taken on the smartphone. The tweets therefore gave participants the opportunity to explore the interplay between text and image as they progressed along the path of the canal, a simple yet powerful augmentation of their creativity. Each tweet became part of a single, collective narrative, a real-time amalgamation of twenty stories that each participant could access alongside their own individual record through the project hashtag or geolocationary tagging.

In the written feedback, one participant called this 'cataloguing the joint experience via social media'. It was this 'joint experience' that seemed to be a key part of the event's popularity. When asked what they enjoyed about the event, over half of the participants stressed its communality: 'companionship of other walkers'; 'meeting new people' and 'enjoyable group experience'. Yet, for almost all the participants, the event's success had also been the way it had encouraged them to see the city in a new way: 'looking at the city in a different light'; 'seeing things that I've passed every day but not noticed' and, perhaps more interestingly, 'going back in time'.

We've already seen how the 'Streets Museum' project, Salford, consisted of a physical installation which travelled around the city (Cassidy). As Cassidy observes, his work was explicitly aimed at 'reintegrating excluded communities, which are often silenced in official sites of memory' (2015, 183). Elissa Rosenberg notes that places remain continually haunted by past meanings and artefacts (2012, 131). In her study of the Berlin memorial project, *Places of Remembrance in the Bavarian Quarter*, Rosenberg describes how Renata Stih and Frieder Schnock's serial installation of street signs deliberately sought to surprise and shock those moving through the city. Rosenberg notes how the project turned walking into a 'transformative encounter' (2012, 132) through which both the meaning of place and the concept of memorialisation itself was challenged.

It was these two ideas – participatory mapping and transformative encounters – that were key to the development of *Waterways and Walkways*. In their analysis of mixed media performance, Steve Benford and Gabriella Giannachi develop the notion of *trajectories* to help conceptualise this kind of interplay between physical and digital elements.

Figure 7.2 Paddle post in the subway under the A470, Cardiff UK.
Source: © Author.

> As lines or threads through experience, trajectories emphasize the journey more than the destination and naturally reflect the importance of continuity and the interweaving of social experience.
>
> (2011, 266)

They define three types of trajectory: canonical, participant and historic (2011, 54). Canonical trajectories defines the artist's intended pathway through a work; participant trajectories describes the actual pathways, or journeys, undertaken by those experiencing the work as users or participants; and lastly, historic trajectories represents the synthesis of what actually took place 'as the experience is subsequently replayed' (2011, 267). For the social media narratives we've explored here, the overt interplay between canonical and participant trajectories seems a particularly useful way of unpacking the overall experience. After all, these social media interventions have many of the characteristics of the sort of mixed reality performances that Benford and Giannachi make the focus of their study. In fact, this tension between the design of the writer/author, and the experience of the user/reader/participant has been a key one that has followed us right across the chapters of this book. From Michael Joyce's concept of constructive (as opposed to exploratory) hypertexts, George Landow's wreader, to the notion of intimate publics and participatory mapping, it has been made clear that postdigital storytelling offers a radicalisation of the creative relationship between author and reader. The stories we've looked at both here and in the previous chapters are, at least in one respect, an exploration of this newfound entanglement.

The outcomes from *Waterways and Walkways* offer at least one model by which the distinct characteristics of social media can be utilised by practice-led research interventions. The collaborative, dialogic, emergent, personalised and context-rich aspects highlighted by Page were all central to the success of *Waterways and Walkways*. Through the hashtag #GlamCan, individual tweets formed a collective, shareable story that was composed in real time as participants walked through the city. The geotagging ensured that each tweet became, what Benford and Giannachi call 'contextual footprints' (20), in other words, digital traces that last beyond the life of the experience, and that also offer the potential of changing future behaviour (21).

Undoubtedly there are similarities with the Situationist International's proposed (although never undertaken) dérive through Amsterdam discussed in Chapter 6, in which the participants' movements would have been controlled by a director using a walkie-talkie from Amsterdam's Stedelijk Museum ('Die Welt als Labyrinth' 1960, 6). Certainly psychogeography's sense of gaming, of playful exploration of urban space, was a significant influence on *Waterways and Walkways*. In her book on radical game design, Mary Flanagan uses the term 'critical gaming' to describe a form of gaming which overtly addresses wider questions of human existence (2009, 6). As Flanagan notes, '[c]ritical play is characterized by a careful examination of social, cultural, political, or even personal themes that function as alternates to popular play spaces' (2009, 6). *Waterways and Walkways* explicitly engaged with a number of issues, including our relationship to urban space and each other at a time of increasing anxiety about the resilience of urban co-habitation (see the United Nation's *Transforming our World: The 2030 Agenda for Sustainable Development*). The creative responses captured by Twitter during the navigation of the lost Glamorganshire Canal highlight the importance that the past can play in any transgressive spatial practice. As Phillips and Milner state in their discussion of online collective storytelling, 'straightforward demarcations between author and audience, between this text and that text, between universal meaning and audience-specific meaning' soon begin to crumble (2017, 161). If cities are, at least in part, the stories we tell ourselves and the itineraries we take across the cityscape, then projects such as *Waterways and Walkways* point to a more radical understanding of participation than that posited by the Centre for Digital Storytelling for example, embracing new ways of storytelling, new ways of making sense of the increasingly contested space that forms our home. As Karen Till notes on the politics of memory in Berlin, 'places are not only continuously interpreted; they are haunted by past structures of meaning and material presences from other times and lives' (2005, 9). Writers too, such as W. G. Sebald and Iain Sinclair, make us aware that the past is never truly expunged from the high street but rather lurks ghost-like in its shadows. In this sense we are always amongst the ruins of what has gone before, both physical space but also human memory. *Waterways and Walkways* suggests that it is social media's ability to reconnect these two phenomena through what I have termed *embodied space*, to remap memory with the lost physical places of our cityscapes through story, that remains one of its strongest features. Yet projects such as *Waterways and Walkways* also remind us that social media is fundamentally hybridic, a postdigital interplay of online/offline, the embodied and the physical. What we've examined in both this and the previous chapter are interventions that have explicitly sought to utilise this phenomenon as part of the storytelling process. Each geotagged tweet of *Waterways and Walkways* was a micro-exploration of embodied space, the wilful narrative interplay of textual and physical worlds.

Novels and social media

As we've seen in Chapter 4, storytelling has always been intimately connected to technological developments. These developments have not only radically changed the way books are published and read; just as importantly, they have impacted on the creative expression of artists themselves. Félix Fénéon's three-line fait divers in *Le Matin* were certainly the outcome of changes in late nineteenth-, early twentieth-century, newspaper publication. Yet they also reflected much wider social changes in reading habits, taste and cultural aesthetics too.

The same is equally true today. Only now, the printed work is no longer the dominant form. As I've argued throughout this book, recourse to the printed page is increasingly an overt and strategic act or artistic intervention. In our postdigital age, the printed novel is no longer a ubiquitous, universalised norm. Instead, it exists in a kind of sublimated tension, or friction, with digitality. To read a printed work of fiction is not only to read a story; it is also to self-consciously step into a non-digital subordinate domain, a domain of printed words, of physical pages and book covers. This transgressive movement *across*, from digital to non-digital, is as much a part of the affective storytelling process as is the textual narrative itself.

For some authors, the non-digital book becomes a perfect way of othering the forms and conventions of the dominant digital domain, particularly the pervasive effects of social media. Joanna Walsh's novel, *Break.up* (2018), for example, traces the effects of a failed relationship on a female protagonist. The relationship had been conducted entirely online and the novel is as much about our relationship to digital technology, particularly social media, as it is about love and loss. ' "I am digital," I said, "not analogue. On or off, zeros and ones" ' (2018, 7). As autofiction, *Break.up* deliberately confuses the boundary between fiction and non-fiction. As the narrator travels from city to city, we get chapters that resemble blog posts, intimate and emergent, self-reflective with their amateurish black and white photographs and chronological time stamp. The breakup of the title refers to the physical relationship; yet it also references the fractured, atomising effects that social media can have on our lives.

In Zadie Smith's novel, *NW*, the large central chapter called 'Host', consists of 185 micro-chapters, numbered sequentially with their own title. Most are no more than 200–300 words in length. Some are longer, some much shorter. For example, here is one of the shortest:

> *38. On the Other Hand*
> Beggars cannot be choosers.
>
> (2012, 192)

The effect of these sections is to provide a kind of amalgam of voice, story and place through which the backstory of Natalie Blake emerges. They are written in omniscient third-person, so in that sense they have a very different tone and register to normal social media entries. However, the structure of this entire section of the novel is striking. The linearity of the entries, their intimate, confessional nature, together with the heavy references to digital technology in later sections (such as the online conversation in chapter 123) inevitably reaches out to the online forms and conventions of social media, in what Wendy Knepper has called Smith's 'postmillennial experimentation' (2013, 111).

In Jon McGregor's *Reservoir 13* (2017), this formal experimentation becomes a central element of the entire narrative. The novel is about the disappearance of a young girl; the story is told over thirteen chapters, each spanning a single year. Yet it is the narrative voice and structure that is most intriguing. The story is told through an unidentified passive voice; each chapter consists of separate sections; individual sentences are written as self-contained, or independent, declarative statements. It's as though we are reading a concatenation of fait-divers or Cole's small fates. Repetition and a neutral register ameliorate the sudden and disorientating switches in focus and subject as the reader moves from sentence to sentence. Several reviewers have stated that these formal effects allow McGregor to replicate village gossip or localised myth-making (see Crown 2017). Yet, I would argue they can also be taken as a structural response to the way social media increasingly dominates personal communication. As we've seen, Hinton and Hjorth use the term 'intimate publics' (2013, 44) to describe this relatively new phenomenon; what we see in *Reservoir 13* is a kind of creative abstraction of that public (or publics), its collaborativeness, fragmentedness and disorientation, the way it enmeshes, embeds, repeats and juxtapositions. These are all characteristics of the formal structuring of the Twittersphere, of course. In that sense, *Reservoir 13*, just like Walsh's *Break.up*, or Teju Cole's small fates project, is as much about how and in what manner we tell our stories in our postdigital world, as it is about the story itself.

Summary

Much has been made about the harmful effects of social media. Significant international events such as the rise of new forms of political populism have been laid at the door of digital platforms such as Facebook and Twitter. Spurred on by growing levels of inequality and the perceived injustices of immigration and globalisation, the world seems suddenly to be filled with hate and anger. Terms such as post truth and fake news speak of a new kind of scepticism, an anti-intellectualism in which any opposing opinion, no matter how grounded in fact, can be written off as propaganda. It seems, then, that while we are more connected than at any time in human history, communal spaces for argument and debate are collapsing; the curation of common ground in which people can reach out to each other is becoming harder. Empathy and compassion are being replaced by levels of ideological dogmatism not seen since the 1930s.

From this perspective, the collaborative, dialogic, emergent, personalised and context-rich aspects of social media are key villains, helping to establish post-truth echo chambers in which people see and read nothing but like-minded opinion and views. For some academics this is reason enough to leave social media well alone. In *Platform Capitalism* (2017), Nick Srnicek argues that emerging technologies such as social media are 'implicated within a system of exploitation, exclusion, and competition' (2017, 7). In this he echoes the views of David Golumbia, who, as we've seen in Chapter 1, used the term "computationalism' to describe the deep socio-cultural enmeshing of technology.

What I have argued in this chapter, however, is that one shouldn't simply right off social media as a pariah technology. While there remain undoubted issues regarding these platforms, both in terms of their ownership and use, I have tried to indicate the ways in which these technologies can be used beneficially within an arts and humanities research context. As *The Economist* notes about smartphones more generally:

> Yes they can be used for wasting time and spreading disinformation. But the good far outweighs the bad. *They might be the most effective tool of development in existence.*
>
> (2019, 14, my italics)

In essence, then, what I've attempted to show here is that some of those features highlighted by Page have real and tangible benefits in terms of storytelling. In Chapter 1, I discussed Alexander Galloway's work on François Laruelle (2014); Galloway is explicit in his conclusion that the movement towards freedom, as he terms it, can only come from within the 'sociopolitical sphere' (2014,

81). Such 'resistance against' must therefore come from within the dominant modality of postdigitality. From meme-campaigns such as #ohneMauerfall or #metoo, to more complex interventions such as *Waterways and Walkways* or *I Tweet Dead People*, social media has been shown to offer a way of encouraging people to 'think differently', as Facer and Pahl describe it (2017, 226). The narrative or storytelling capability of the technology is central to this transformative experience. Using Twitter as a case study, I have shown how the collaborative, emergent and highly personalised aspects of social media can become real strengths in terms of the co-creation of knowledge. The fact that social media is mobile and emerges dialogically, in real time, were key to its success in these projects.

What's also been central to this discussion is an understanding of social media as inherently entangled, embodied and hybridic. Platforms such as Twitter, Facebook and Instagram, are far less straightforwardly 'digital' than the PC-bound, dial-up enabled, discussion boards and forums of the late 1990s and early noughties. Mobile phone technology has given social media vastly increased opportunity for new forms of entanglement. In the previous chapter I introduced one such innovation, a new form of narrative space, *embodied space*, in which there was a deliberate entanglement of textual and embodied worlds. Social media can be seen as offering further interventions into this new hybridic domain. Yet unlike the locative stories discussed in the previous chapter, the storytelling we've seen here has been inherently collaborative, emergent and dialogic. In other words, as a mode of postdigital storytelling, social media is even more overtly autofictional than the locative mobile stories discussed in Chapter 6 (Gibbons 2017). Here, the emphasis is on a new kind of affective storytelling, a post-ironic metamodernist sensibility, in which empathy and social connectedness are purposely foregrounded (see Elizabeth A. Segal *Social Empathy: The Art of Understanding Others* 2018). While this underpins all the research discussed in this chapter, including the CDS' model of collaborative storytelling with its emphasis on co-creation and mutuality, it also inflects non-digital works too. As we've seen, novels such as Zadie Smith's *NW*, Jon McGregor's *Reservoir 13* and Joanna Walsh's *Break.up*, as well as being non-digital responses to social media, can also be interpreted as works seeking new forms of truthmaking, a new language, if you will, in which the subjectivity of perception is balanced against an imperative for affectivity and meaning.

In these ways, social media resembles a kind of hybridic weft or weave that is always emergent and very often collaborative. Text derives from the Latin *textus*, or 'woven' (Ingold 2016, 63); from this perspective, to send a text is to create a thread. In *On Weaving* (2017), the textile artist Anni Albers uses the term 'the event of a thread' to describe both the singularity of a single thread as well as its interaction with other countless threads, a multilinear relationship without beginning or end (xi). This open-ended interplay of threads is mediated through the relationship between weaver and machine. Ingold's term meshwork seems equally relevant here; from this perspective, both the construction of a text and the pulling of a thread are a form of wayfaring – the 'movement along', or Albers' event – through which our being-in-the-world is instantiated.

As a form of postdigital wayfaring, then, social media offers a powerful augmentation of the locative storytelling we've explored in the previous chapter, a narrativising that is fundamentally hybridic, emergent and collaborative. If a tweet or Facebook post is a single thread, then the tapestry is what is created by the limitless opportunity for digital and non-digital enmeshment, the interlinking and contextual, situated embedding of embodied space.

I've been using the metaphor of the string-figure to represent the new creative modality of our post-postmodern era. As I explained in Chapter 2, for Haraway, the string-figure is a powerful metaphor for a new kind of interconnected storytelling – individual and collective, shared and co-created – that is capable of engaging with and, ultimately, changing the world. This 'making trouble', as Haraway calls it, has been an important aspect of the projects examined in this chapter too, including those that have drawn on the collaborative ethos of the CDS' methodology. What I've also highlighted is that social media, as a key facet of postdigital storytelling, can play a powerful role in the creation of affective and embodied stories. Not so much intimate public then, but publics; a dialogic interplay of 'troublesome' knowledge (Meyer and Land 2005, 373–374). From this perspective, social media is not necessarily straightforwardly exploitative as a technology (Srnicek 2017). Phillips and Milner (2017) use the adjective ambivalent; although not without its own complications too, the term is useful in the way it brings the embodied experience of the user back into the debate. Yet whatever position one takes, it is clear that social media is not going to disappear anytime soon. While it is important to recognise the attendant issues and controversies associated with the technology (just as it is necessary to do for any media), I hope this chapter has made some headway in explaining why academia would be unwise to reject social media out of hand.

Notes

1 See www.storycenter.org/storycenter-blog/blog/2013/3/4/wynnes-story.html.
2 A good example of this is the Facebook-Cambridge Analytica data scandal in 2018. See Carole Cadwalladr and Emma Graham-Harrison, 'Revealed: 50 million Facebook profiles harvested for Cambridge Analytica in major data breach', *The Guardian*, 2018.
3 We already examined early discussion boards and forums in relation to *The House of Leaves* (2000) in Chapter 5.
4 See www.newyorker.com/magazine/2012/06/04/black-box-2.
5 See the project website https://waterandwalk.wordpress.com.

Works cited

Albers, A. *On Weaving*. Princeton: Princeton University Press, 2017 [1965].
Anderson, S. *Technologies of History: Visual Media and the Eccentricity of the Past*. Boston: Dartmouth College Press, 2011.
Andrejevic, M. 'Social Network Exploitation'. *A Networked Self: Identity, Community, and Culture on Social Network Sites*. Ed, Paparcharissi, Z. New York: Routledge, 2011, 82–101.

Anon., 'Die Welt als Labyrinth'. *Internationale Situationniste*. 4 June 1960: 5–7.
Attlee, J. *The Cartographer's Confession*. James Atlee, 2018.
Benford, S. and Giannachi, G. *Performing Mixed Reality*. Cambridge, Massachusetts: MIT Press, 2011.
Bernstein, J. H. 'Transdisciplinarity: A review of its origins, development, and current issues'. Journal of Research Practice, 11 (1), 2015. Web. 15 February 2018. http://jrp.icaap.org/index.php/jrp/article/view/510/412.
Bolt, B. *Heidegger Reframed*. London: I.B. Tauris, 2011.
Cadwalladr, C. and Graham-Harrison, E. 'Revealed: 50 Million Facebook Profiles Harvested for Cambridge Analytica in Major Data Breach'. *The Guardian*, 17 March 2018. Web. 5 November 2018. www.theguardian.com/news/2018/mar/17/cambridge-analytica-facebook-influence-us-election.
Cassidy, L. 'Salford 7/District Six. The Use of Participatory Mapping and Material Artefacts in Cultural Memory Projects'. *Mapping Cultures: Place, Practice, Performance*. Ed, Roberts, L. London: Palgrave Macmillan, 2015. 181–198.
Cole, T. 'Small Fates'. March 2011. Web. 21 November 2018. www.tejucole.com/other-words/small-fates.
Crown, S. 'The Rhythms of Rural England in Jon McGregor's Novel of Forensic Noticing'. *Times Literary Supplement*, 29 March 2017. Web. 24 November 2018. www.the-tls.co.uk/articles/public/jon-mcgregor.
D'Ancona, M. *Post-Truth: The New War on Truth and How to Fight Back* London: Ebury Press, 2017.
Dunford, M. and Jenkins, T. 'Form and Content in Digital Storytelling'. *Digital Storytelling: Form and Content*. Eds, Dunford, M. and Jenkins, T. London: Palgrave Macmillan, 2017. 1–17.
The Economist. 'Bad News for Apple. Good News for Humanity'. *The Economist*, 12–18 January 2019, 14.
Egan, J. 'Black Box'. *The New Yorker*. 4 June 2012. Web. 21 November 2018. www.newyorker.com/magazine/2012/06/04/black-box-2.
Facer, K. and Pahl, K. 'Understanding Collaborative Research Practices: A Lexicon'. *Valuing Interdisciplinary Collaborative Research: Beyond Impact*. Eds, Keri Facer and Kate Pahl. Bristol: Polity Press, 2017, 215–231.
Fénéon, F. *Novels in Three Lines*. New York: New York Review of Books, 2007.
Flanagan, M. *Critical Play: Radical Game Design*. Cambridge, MA: Massachusetts Institute of Technology, 2009.
Flood, A. 'Novelist David Mitchell Publishes New Short Story on Twitter'. *The Guardian*. Monday 14 July 2014. Web. 21 November 2018. www.theguardian.com/books/2014/jul/14/david-mitchell-publishes-short-story-twitter.
Galloway, A. R. *Laruelle: Against the Digital*. Minneapolis: University of Minneapolis Press, 2014.
Gibbons, A. 'Contemporary Autofiction and Metamodern Affect'. *Metamodernism: Historicity, Affect, and Depth After Postmodernism*. Eds, van den Akker, R. Gibbons, A. and Vermeulen, T. London: Rowman & Littlefield International, 2017, 117–130.
Golumbia, D. *The Cultural Logic of Computation*. Cambridge, Mass.: Harvard University Press, 2009.
Hallett, C. W. *Minimalism and the Short Story: Raymond Carver, Amy Hempel, and Mary Robison*. New York: Edwin Mellen Press, 1999.
Haraway, D. J. *Staying with the Trouble: Making Kin in the Chthulucene*. Durham: Duke University Press, 2016.

Hessler, B. and Lambert, J. 'Threshold Concepts in Digital Storytelling: Naming What We Know About Storywork'. *Digital storytelling in Higher Education: International Perspectives*. Eds, Jamissen, G., Hardy, P., Nordkvelle Y. and Pleasants, H. London: Palgrave Macmillan, 2017, 19–35.

Hinton, S. and Hjorth, L. *Understanding Social Media*. London: Sage Publications, 2013.

Ingold, T. *Lines: A Brief History*. London: Routledge, 2016.

Jenkins, H. *Convergence Culture: Where Old and New Media Intersect*. New York: New York University Press, 2006.

Joyce, M. 'Siren Shapes: Exploratory and Constructive Hypertexts'. *Of Two Minds: Hypertext, Pedagogy and Poetics*. Ann Arbor: University of Michigan Press, 1995 [1993], 39–59.

Kemp. S. *Digital in 2018: Essential Insights into Internet, Social Media, Mobile, and Ecommerce use Around the World*. London: We Are Social/Hootsuite, 2018.

Knepper, W. 'Revisionary Modernism and Postmillennial Experimentation in Zadie Smith's *NW*'. *Reading Zadie Smith: The First Decade and Beyond*. Eds, Tew, P. London, Bloomsbury, 2013, 111–126.

Lambert, J. *Digital Storytelling: Capturing Lives, Creating Community*. London: Routledge, 2013.

Landow, G. P. *Hypertext: The Convergence of Contemporary Critical Theory and Technology*. Baltimore: John Hopkins University, 1992.

McGregor, J. *Reservoir 13*. London: 4th Estate, 2017.

Maggi, W. Participant-Observation. Berkeley, CA: Centre for Digital Storytelling, 2011. Web. 4 November 2018. www.storycenter.org/storycenter-blog/blog/2013/3/4/wynnes-story.html.

Meyer, J. H. F. and Land, R. 'Threshold Concepts and Troublesome Knowledge (2): Epistemological Considerations and a Conceptual Framework for Teaching and Learning'. *Higher Education*, 49 (3), 2005, 373–388.

Misorelli, C. 'One Million Life Stories'. *Digital Storytelling: Form and Content*. Eds, Dunford, M. and Jenkins, T. London: Palgrave Macmillan, 2017, 35–39.

Montuori, A. 'Foreword: Transdisciplinarity'. *Transdisciplinarity: Theory and Practice* Ed, Nicolescu, B. C., New Jersey: Hampton Press Inc.: 2008, ix–xvii.

Ofcom. *Adults' Media Use and Attitudes Report*. London: Office of Communications, 2018.

Page, R. E. *Stories and Social Media: Identities and Interaction*. London: Routledge, 2012.

Phillips, W. and Milner, R. M. *The Ambivalent Internet: Mischief, Oddity and Antagonism Online*. Cambridge: Polity Press, 2017.

Rosenberg, E. 'Walking in the City: Memory and Place.'. *The Journal of Architecture*, 17 (1) 2012, 131–149.

Sante, L. 'Introduction' in Fénéon, F. *Novels in Three Lines*. New York: New York Review of Books, 2007, vii–xxxi.

Segal, E. A. *Social Empathy: The Art of Understanding Others*. New York: Columbia University Press, 2018.

Skains, R. L. *Digital Authorship: Publishing in the Attention Economy*. Cambridge: Cambridge University Press, 2019.

Smith, Z. *NW*. London: Hamish Hamilton, 2012.

Smyth, K., Power, A. and Martin, R. 'Culturally Mapping Legacies of Collaborative Heritage Projects'. *Valuing Interdisciplinary Collaborative Research: Beyond Impact*. Eds, Facer, K. and Pahl, K. Bristol: Polity Press, 2017, 191–213.

Speakman, D. *It Must Have Been Dark By Then*. Bristol: Taylor Brothers, 2017.

Speed, C. 'Walking Through Time: Use of Locative Media to Explore Historical Maps'. *Mapping Cultures: Place, Practice, Performance*. Ed, Roberts, L. London: Palgrave Macmillan, 2015, 160–180.

Srnicek, N. *Platform Capitalism*. Cambridge: Polity Press, 2017.

Till, K. *The New Berlin: Memory, Politics and Place*. Minneapolis: University of Minnesota Press, 2005.

Tringham, R. 'Creating Narratives of the Past as Recombinant Histories'. *Subjects and Narratives in Archaeology*. Eds, van Dyke, R. M. and Bernbeck, R. Boulder. Colorado: University Press of Colorado, 2015. 27–54.

United Nations. *Transforming our World: The 2030 Agenda for Sustainable Development*. New York: United Nations, 2015.

Walsh, J. *Break.up*. London: Tuskar Rock Press, 2018.

Walsh, J. *Seed*. London: Editions At Play, 2017.

Worcman, K. and Garde-Hansen, J. *Social Memory Technology: Theory, Practice, Action*. London: Routledge, 2016.

Part 3

Coda

8 How soon is now?

The American artist and author, Trevor Paglen, has spent much of his professional career exposing and interrogating the hidden world of technologically-enabled state surveillance, monitoring and control. In his project *How to See Like a Machine* (2017), Paglen attempted to give visual form to the hidden communication that occurs within and between machines. The context for these works is what Paglen identifies as the increasing proliferation of 'machine vision' and artificial intelligence (AI) in which 'images are made by machines for other machines, with humans rarely in the loop' (Cornell *et al.* 2018, 33).

Paglen and his team developed software that allowed them both to replicate automatic image production and recognition algorithms, and then create visual images of the data generated by those algorithms. In essence, Paglen's software was an attempt at allowing us to see what the machines were seeing – to see like a machine. Paglen processed the images of a number of political philosophers, including the radical postcolonial thinker, Frantz Fanon. *Fanon (Even the Dead Are Not Safe) Eigenface*, shown in Figure 8.1, is a dye sublimation metal print; eigenface refers to the algorithmic modelling used in human-face recognition software.

Related work, such as *Shark, Linear Classifier* (2017a), uses an object recognition technique called linear classifier to produce a composite image: instead of a shark, Paglen's image resembles an abstract watercolour of blues and whites. Yet, as he notes, 'where the *Shark* image looks indistinct and abstract to human eyes, it's an exceptionally specific representation to the computer vision system' (Cornell *et al.* 2018, 35). *Shoshone Falls, Hough Transform; Haar* (2017b) portrays how two image analysis algorithms (known as Hough Transform and Haar) visualise or 'see' a photograph of the famous Idaho falls. This time the subject is less abstract, more recognisable as a torrent of water, despite the disconcerting overlay of geometric lines and shapes.

The overall effect of these and other, related, work is at least twofold: first, they draw attention to the deepening gaze that AI has on each and everyone one of us, whether it be through face recognition software, automated license plate readers or drone image capture. As Cornell, Bryan-Wilson and Kholeif note, '[w]e are increasingly uncertain about what might be looking back at us, how we are seen, and for what purpose' (2018, 101). Secondly, and perhaps

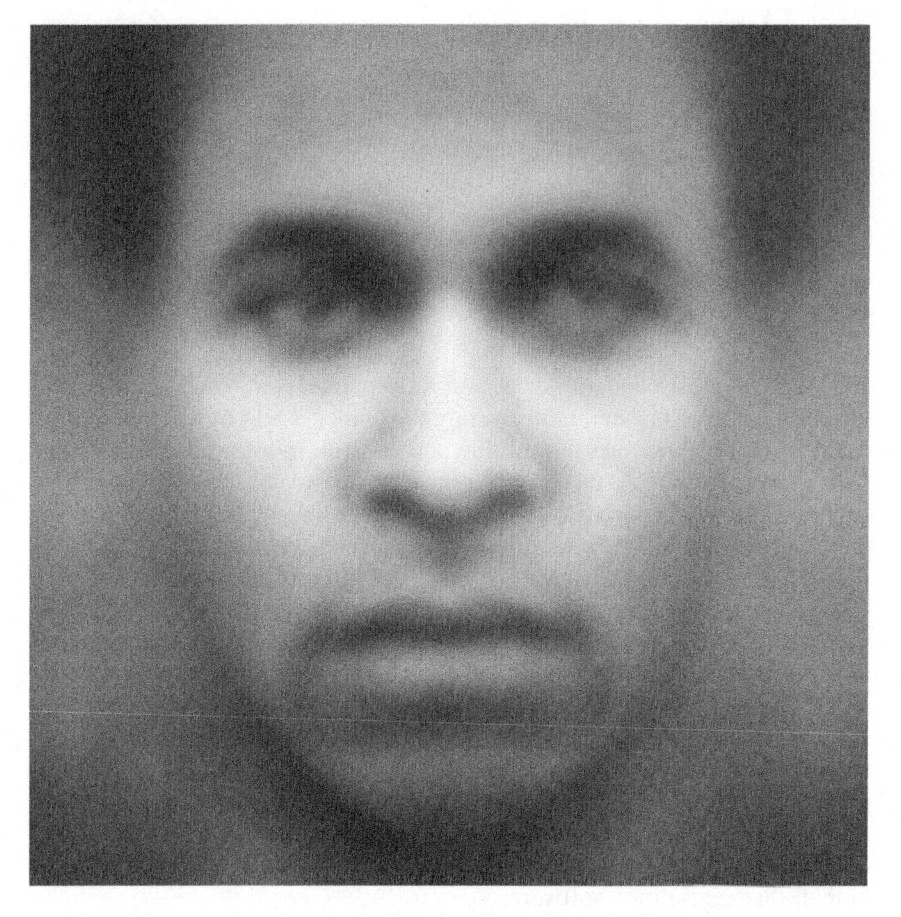

Figure 8.1 Fanon (Even the Dead Are Not Safe) Eigenface by Trevor Paglen, 2017. Dye Sublimation Print 122 × 122 cm.

Source: © Trevor Paglen. Courtesy of the artist and Metro Pictures, New York.

more importantly in regard to the focus of this book, they foreground a way of thinking about machine-based logic that is not straightforwardly computational or code-driven. Much of Paglen's work seeks to expose the invisible influence of, what Dwight Eisenhower termed, the military-industrial complex; *How to See Like a Machine* certainly does this, yet it does something else too. Paglen's concept of 'machine vision' carries with it notions of cybernetic consciousness, of an inscrutable aesthetic that, though serving government or corporate aims, will nevertheless always lay beyond human comprehension. In Paglen's hands, a visual artist, this aesthetic is very much image based; yet it would be just as relevant to think of this hidden substrata of machine-to-machine 'seeing' as cybernetic conversations or even stories. If that is the case, then what Paglen and his team of programmers have rendered visually is a form, perhaps the purest form, of digital storytelling, a storifying in which embodied human input is erased. From this perspective, works such as *Fanon (Even the Dead Are Not Safe) Eigenface* are images of a possible future in which the stories we tell of each other and the world are increasingly digital-to-digital, an endless stream of data generated through the impenetrable gaze of AI. In this world, digital storytelling becomes entirely cybernetic, hidden away in the encoded nucleus of my code/data/narrative/materiality model outlined in Chapter 3.

This conceptualisation may seem fanciful, or even hysterical, to some; nevertheless I feel it offers a powerful way of re-engaging with what I've been arguing throughout this book. At the heart of my discussion about digital and postdigital storytelling has been the central issue of human agency. Right from its earliest incarnation, the term postdigital has been used to describe the need for humans to somehow reinscribe themselves back onto the digital artefact. As Hannah Arendt states, '[n]o human life, not even the life of the hermit in nature's wilderness, is possible without a world which directly or indirectly testifies to the presence of other human beings' (1998, 22). The distortion and glitches that Kim Casone observed on digital recordings at the end of the last century were exactly that – early attempts by artists at rebalancing the creative interplay between body and machine. And what better definition is there of this basic human desire to hack back into the cybernetic systems they have created than Florian Cramer's 'post-digital hacker attitude' (2015, 20).

In the world of AI, drones, and surveillance satellites, postdigital takes on radical new potential, drawing out imperatives and concerns that lay hidden within those early conceptualisations of the term. While posthuman offers a means of rethinking human subjectivity in a technologicalised, globalised context (Braidotti 2013), postdigital turns the focus back onto the technological domain itself, and our interaction with it. In this way, the concept of postdigitality is far closer to the radical intent of artistic work such as Paglen's *How to See Like a Machine*. I would argue that *Fanon (Even the Dead Are Not Safe) Eigenface* (2017) and *Shark Linear Classifier* (2017) invite less a reflection on posthuman subjectivity, than an imperative for intervention in terms of how we, as humans, physically and cognitively interact with computer systems.

The postdigital paradigm holds human subjectivity fixed or steady, while positing change in how we engage with the digital domain; posthumanism takes the reverse position, holding fixed our interaction with a technologicalised domain, while concurrently foregrounding the impact on human subjectivity that such a domain entails. This is not to say that postdigital does not embrace subjectivity; on the contrary, as I've shown throughout this book, human experience as a situated and embodied experience lies right at the heart of any understanding of the concept. Rather, I would say that postdigital has as its emphasis the potentiality, and indeed the imperative, of human agency *on* digital systems. In that sense, the focus of postdigitality is on the sort of worlding of the world, to use Heidegger's term, that only comes through practice-based intervention. If posthumanism betrays its postmodernist heritage, problematising Enlightenment concepts of the self, then postdigitality draws on the very different ontology of metamodernism. This has been a key argument of this book.

Postdigital/storytelling

It would be tempting to say that this book begins and ends with a selfie. Yet that wouldn't be quite true. Instead, what we have are two images that capture, in their own very different ways, the selfie as both a social and technological phenomenon. Barbara Kinney's photograph of Hillary Clinton, shown in Figure 1.1, became an internet sensation in the way it was seen to encapsulate the habitus of the millennial generation. Whereas previous generations would have taken a photograph of the presidential candidate, Kinney's image records a new social urge to inscribe oneself into the moment. In this sense, what we see in Figure 1.1 is not the straightforward desire to simply record the external world; rather, what Kinney caught in her photograph is the sociological imperative for new forms of situated embodiedness in an increasingly complex and fragile world. From this perspective the selfie is exactly that – a technologically-mediated inscription of the self into the world, a hybridic storifying conjured up across the emergent and dialogic structures of social media.

The second image is more complex. *Fanon (Even the Dead Are Not Safe) Eigenface*, shown in Figure 8.1, appears to be a selfie but of course isn't. Instead, what we have is a kind of encrypted selfie, if you will, in which the sensorium of a digital system is represented. The selfie component (the selfieness) of this image is therefore not represented by the *content* (a portrait of Frantz Fanon) but rather is contained in the unworldly rendering of its *form*. If the form and content of the selfie in Figure 1.1 combine to say 'This is you', then the selfie in Figure 8.1 can ultimately only say one thing to us: 'This is not you'.

These two images capture the tension that has been at the heart of this book: the form and function of human agency in an increasingly technologicalised world. The human agency I've been looking at, of course, is storytelling, the imperative for humans to tell stories and the form that those stories take. Artists such as Paglen remind us, however, that we are fast approaching a time when the majority of machine activity will not involve any direct human input. It is this new form of digital storytelling that we see staring back at us in Figure 8.1.

Postdigital storytelling, then, becomes nothing less than the imperative to tell and retell the 'being-in-the-world' of the human condition. Critically, however, I've contextualised this in two ways. The first is that I've positioned postdigital storytelling as a key mode or characteristic of a new ontological paradigm. As the ironic scepticism of postmodernism has fallen away, it has been replaced by a desire for new forms of moral and ethical connectedness, what Robin van den Akker and Timotheus Vermeulen describe as 'the return to realist and modernist forms, techniques and aspirations' (2017, 3). This new dominant cultural logic, what some have labelled metamodernism, a term I adopt in this book, has brought with it a paradigm shift in artistic representation. In his overview of Western European artistic representation, Richard Kearney isolated three significant periods, or paradigms, in which each privileges a separate 'metaphor characterizing the dominant function of imagination' at that time (1988, 17): he labelled these mirror, lamp and labyrinth. The last of these was the postmodern labyrinth of the looking glass, reflecting infinite variations on an ultimately illusory object. What I've offered in this book is the notion of the map/rhizome/string-figure as the fourth creative modality within Kearney's schematic. Taken together, I've argued that the map, the rhizome and string-figure provide a powerful set of 'images of imagining' by which contemporary, affective, creativity can be both visualised and at the same time demarcated from postmodernism. The map, the rhizome and string-figure are powerful metaphors for metamodernist creativity or wayfinding, foregrounding the sort of situated and embodied approaches espoused by van den Akker, Gibbons and Vermeulen. In this new era of metamodernity, the map, rhizome and string-figure are reconfigured as symbols of our subjectively intimate and ethically relational connectedness.

Secondly, I've approached postdigital storytelling as a form of academic research. Martin Heidegger's concept of 'praxical knowledge' has been critical here, allowing the foregrounding of storytelling as a primary means of knowledge creation (Bolt 2011). Central to this has been to rethink what creativity might actually mean, both from a metamodernist perspective, and from the perspective of academic research and praxis.

> Creativity is a situated and embodied act, in which, through the (co-) creation of new knowledge, perceptions and understanding of the world are changed.

This definition has been key to my critical exploration of postdigital storytelling. A reworking of Rob Pope's definition, it places emphasis on three important themes: an understanding of creativity as a fundamental process of *knowledge creation*; an understanding of creativity as an inherently *situated* and *embodied* activity; and finally, an understanding of creativity as a means of '*becoming other-wise*', to use Pope's term (2005, 29). This third point is perhaps worth stressing: creativity as practiced within academia cannot afford to be solipsistic. Rather, it must seek to affect change, to cause trouble, as Donna Haraway would have it. Yet all three themes clearly signal the metamodernist sensibilities of my definition.

If we are to critically engage with creative practice today, then we need to push beyond traditional concepts. Digital storytelling enshrines a conceptual binarity of the digital and non-digital domains that, as I've argued throughout this book, no longer holds. There was a time, of course, when thinking of something as distinctly digital, as opposed to non-digital, made sense. This was certainly the case throughout much of the last quarter of the twentieth century. Here, new and innovative forms of digital technology combined with the overarching postmodern paradigm of those years to nullify any understanding of the self as coherent and stable. Digital works such as Michael Joyce's *afternoon: a story* (1987) and Shelley Jackson's *Patchwork Girl* (1995) self-consciously championed the end of print, while George Landow's concept of 'wreader' (1992) beckoned towards a utopia of non-linearity and indeterminacy in which narratives were as much an exploration of their own fictionality as of any overt authorial story. Hypertext authors of the 1980s and early 90s took postmodern novels such as Thomas Pynchon's *Gravity's Rainbow* (1973) and transferred their artistic DNA to the purifying world of the computer screen. For these authors and artists, then, digital technology offered unique and powerful ways to represent something so profound and pervasive that it encompassed every aspect of contemporary life: postmodernity.

By the early 1990s, however, these certainties were beginning to look outdated. It is this moment, this shift in perception, that forms the fulcrum for this book. If we go back to Kearney's schematic, then what we see from this time is the transition away from the postmodern labyrinth towards the modality of, what I've labelled, map/rhizome/string-figure. Hypertextual work such as *The Unknown* (1998) and *The Doll Games* (2001) are some of the earliest examples of this creative modality, in which the desire for an affective experience, embracing both a situated and embodied understanding of storifying is pushed to the fore in a way that previous hypertextual work had not. Concurrently, Mark Z. Danielewski's *House of Leaves* (2000) had pulled apart and then rebuilt the 'classic' postmodern novel. Only now, to use Jessica Pressman's term, it was 'networked' (2006). In other words it was hybridic, the story evolving across a transmedial landscape of physical print, music and digital text. And just like *The Unknown* and *The Doll Games*, *House of Leaves* exhibited a profound focus on authentic emotional affect and situated experience, the sort of stuff that, as Mary K. Holland reminds us, 'postmodernity had long jettisoned' (2013, 98).

If this is the fulcrum of this book, then it is one whose conceptual underpinnings need to be reiterated: any understanding of our post-postmodern condition, the metamodernist modality of the map/rhizome/string-figure, must also be seen as a response to postdigitality. In other words, the critical argument of this book is that postdigitality is an underlying and intrinsic characteristic of metamodernism.

This relationship between metamodernism and the postdigital becomes more overt as the analysis is extended into the twenty-first century. Digital fiction such as *These Pages Fall Like Ash* (2013) by Tom Abba and Duncan Speakman, *Seed* (2017) by Joanna Walsh and *Breathe* (2018) by Kate Pullinger, are stories in which the embodied and situated nature of their readers becomes a central feature of the narrative. Indeed, one could say that the primary objective of these fictions is to extend the narrative interplay between the digital and non-digital domains. The development of locative mobile technology, such as the smartphone, has extended the ability of authors to explore this phenomenon even further. *The Cartographer's Confession* (2018) and *It Must Have Been Dark By Then* (2017) are just two examples of work that explores a new kind of narrative space, what I have called *embodied space*. Embodied space is a hybridic form of narrative space that arises from the text but which also has direct and intentional connections to spaces within the actual or real world. I have argued that it forms a sixth layer in Ryan, Foote and Azaryahu's taxonomy of narrative space and is an important aspect of any notion of postdigital poetics.

What we see in *The Cartographer's Confession* and *It Must Have Been Dark By Then* is a form of postdigital wayfaring, a mediation across digital and non-digital space: a physical journey, at least in part, in the course of which a story evolves. Here, embodied space becomes deeply hybridic, an embodied entanglement of digital and non-digital domains. And as Speakman demonstrates, this hybridity is not necessarily a straightforward one of digital platform and human body. Works such as *It Must Have Been Dark By Then* and *These Pages Fall Like Ash* are early forays towards a more complex form of storytelling, in which an embodied, situated experience is mediated between and across locative mobile technology and physical non-digital texts.

Here then is where my definition of postdigitality moves beyond previous understandings of the term, such as those given by Casone, Cramer, and Berry and Dieter. For Berry and Dieter, for example, concepts of postdigitality are strongly aligned to notions of the *New Aesthetic*, an art movement seeking to explore the wider effects of digital proliferation and convergence. As Christiane Paul and Malcolm Levy state, '[t]he New Aesthetic captures the embeddedness of the digital in the objects, images and structures we encounter on a daily basis and the way we understand ourselves in relation to them' (2015, 27). From this perspective, postdigital takes on two important characteristics. First, it becomes almost exclusively digital in its focus, a phenomenon in which the non-digital is folded into, or subsumed within, the digital. And second, postdigital becomes fundamentally historicised to the point where any understanding of the term as a significant disjuncture is ultimately denied. As Paul and Levy explain in their analysis of the New Aesthetic, it becomes difficult to see the movement as offering 'anything radically new' within this wider, longitudinal perspective (41).

By undertaking an in-depth analysis of postdigital storytelling, primarily fiction, I've been able to challenge these assumptions. Positing postdigitality as a key characteristic of metamodernism (the movement from labyrinth to map/rhizome/string-figure) has allowed me to underscore the degree to which the term marks a significant and decisive break with earlier periods. Not only is this rupture ontological; I've also shown how recent technological developments such as mobile locative technology, social media and AI, should be seen as marking a decisive demarcation within the historical continuum.

Reformulating postdigitality as a response to a changing ontological paradigm has also enabled a broadening of its focus. Previous definitions of postdigital have been very much focussed on the digital domain. With Cramer's 'post-digital hacker attitude' (2015, 20), for example, it is the characteristics of non-digital media that are reinscribed back onto the digital artefact. What I've shown throughout this book, however, is that this movement *across* the domains works in both directions. In other words, postdigitality is as much a condition of the (subordinate) non-digital domain as it is of the (dominant) digital domain. Central to this position is the argument that in an increasingly digitised world, any recourse to non-digitality becomes a strategic decision. Until very recently, the printed book was socially and culturally hegemonic; it was the digital book, the hypertext fiction, that was the interloper. At this time, the digital work was always in deficit, always being defined by that which it was not: the printed book. And crucially, of course, this functional deficit underpinned both the author's and reader's experiences.

In the postdigital age, however, this condition has been reversed. It is now the printed book that is in functional deficit; it is now the printed novel which is defined by that which it is not: the online, interactive, participatory story. For works such as Matthew McIntosh's *theMystery.doc*, Zadie Smith's *NW* and Ben Lerner's *10:04*, it is not just the representation of digital and hypertextual media that is central to their storytelling: critical to their work is the unremitting analogue form of the printed novel itself. If metamodernism is a kind of striving for truth-telling, an ethically-inflected reassertion of the self, then postdigitality is one way of thinking about what that new language might be. As we've seen, it can be hybridic, a mashup of digital and non-digital domains; but it can also be entirely non-digital too, and it's important not to forget this. Printed fiction such as *NW* and *10:04* are equally works of a metamodernist inflection, signifying in their own way a shift in creative modality from the labyrinth of mirrors to the map/rhizome/string-figure. In each of these novels, the exploration of the situated and embodied nature of the author/narrator is just as important as it is in digitally-hybridic works such as *Breathe* and *The Cartographer's Confession*; however, the physicality of these printed works explicitly undercuts the dominant digital sensorium with, what amounts to, new forms of literary representation: a new analogue language, Lerner's notion of 'prosody', by which we may contemplate a postdigital existence.

If digital storytelling is the sort of machine-to-machine interactions captured by Paglen's *How to See Like a Machine* project, then postdigital storytelling describes the narrative interactions of humankind. These postdigital interactions span right across the non-digital/digital spectrum, from the traditional hardcopy book, to the most advanced digital platforms with all kinds of online/offline hybridity in between. Yet, despite this complexity in form, what perhaps has been the most consistent aspect across all these stories has been a re-engagement with authentic emotional affect and situated experience, what Holland, calls a 'reemergence of some sort of earnestness and meaning – an antidote to the destructiveness of [postmodern] ironic detachment' (2013, 123). If there is a central theme to postdigital storytelling then it is surely this: the reaffirmation of our own humanity.

Interdisciplinarity/transdisciplinarity

At a profound level, transdisciplinarity is connection and connectedness.[1] Exploring postdigital storytelling as both a form of creative expression and academic research has been one of the major foci of this book. In one sense, this has given my argument a further degree of refinement: not only am I primarily looking at prose within the wide spectrum of what might be classed as storytelling, but I am also interested in those examples of storifying that are either examples of research in and of themselves, or offer important lessons for practice-based methodologies. Yet, in another sense, my argument about the underpinning metamodernist sensibilities of postdigital storytelling – its foregrounding of affective change, and its ethical and moral inflections – naturally ensures a 'reaching out to' or connection with academic research agendas.

In their review of over three-hundred academic projects funded by the Research Councils UK and the Arts and Humanities Research Council (AHRC), Keri Facer and Kate Pahl list what they see as the key elements of these projects (2017, 17). Even though I've already described these in some detail, they're worth repeating: (1) *Materiality and place* – the fundamental iterative relationship between knowledge production and objects, landscapes and cultures; (2) *Praxis* – knowledge through doing, what Facer and Pahl call a 'performative ontology that shapes the world as it studies it' (2017, 17); (3) *Stories* – the role narrative plays in the creation of knowledge; (4) *Embodied learning* – that knowledge is fundamentally embodied and situated, whether individually or/and through embodied networks and communities; and finally (5) *translation* – the ways in which knowledge crosses borders and thresholds.

Facer and Pahl cite these elements as key characteristics of what they call collaborative research, in other words research that embraces a wide range of participants, including academics, community members and public bodies. The project, *Preserving Place: A Cultural Mapping Exercise*, would be a good example of Facer and Pahl's definition (Smyth *et al.* 2017).

Yet it is also clear that these elements form the foundational attributes of postdigital storytelling. What Facer and Pahl are describing is not simply new modes of academic engagement (collaborative or participatory research) but rather just one aspect of the wider, and deeper, ontological retreat from postmodernism that has been traced out across the chapters of this book. In other words, if metamodernism is, at its heart, the belief in and application of, affective change and connectedness, then, what Facer and Pahl offer us is perhaps best understood as an exploration of the ways in which such change can be theorised.

Facer and Pahl's emphasis on *translation* is particularly telling here; however, its implications go much further than any consideration of collaborative research. The movement of knowledge across socio-cultural boundaries was at the heart of the AHRC's Connected Communities Programme of course; translation captures this sense of different 'languages', of barely interacting knowledge systems, across and between communities and societies. For Facer and Pahl, any engagement with this kind of translation demands an interdisciplinary response, where this movement across boundaries is not only instantiated across different academic disciplines but also between academic and non-academic structures. At one level, this seems a sensible approach. Interdisciplinarity offers an effective conceptual framework by which the affectivity of any practice-based intervention can be considered. Crucially, however, its *modus operandi* is primarily cross-disciplinary; any connection with the non-academic realm is always a secondary action. This unnecessarily limits Facer and Pahl's concept of translation.

This book has argued that transdisciplinarity offers a way forward here. As I argued in Chapter 2, instead of focussing on the sharing of knowledge across academic boundaries, transdisciplinarity 'concerns that which is at once between the disciplines, across the different disciplines, and beyond all disciplines' (Nicolescu 2008, 2). Rosemary Ross Johnston is surely right when she notes that, ultimately, transdisciplinarity is all about connections within, what she terms, an 'ecology of knowledge' (2008, 227). From this perspective, any translation of knowledge is not simply made across and between academic disciplines, but involves a far-more complex interaction of academic and non-academic domains. Transdisciplinarity is therefore not primarily discipline-driven in its focus; rather, it is inquiry-driven, a process in which the overall aim is affective and purposeful change on the world. In that regard, as Alfonso Montuori notes, transdisciplinarity does not prioritise *knowing* per se; rather, it is much more interested in the way such knowledge is constructed in the real world (2008, xi). Within a transdisciplinary context, ideas of translation morph into notions of interaction and connection, foregrounding the central role of human experience, the practice/praxis continuum outlined in Chapter 3. Much more so than interdisciplinarity, then, transdisciplinarity embraces situated embodiedness as a key aspect of knowledge (co)-creation. As a consequence, transdisciplinarity becomes especially pertinent as a metamodernist means of inquiry, in which the emphasis is less on the inviolability of disciplinary knowledge, and much more on the embodied dynamics of affective change.

Postdigital storytelling is a perfect example of this. As Johnston notes, art-based subjects are by their very nature transdisciplinary in that their natural focus is always outwards, beyond the disciplinary boundaries of particular artistic forms and structures:

> The arts offer powerful, transformational, experiential ways of learning; as well as separate subjects (literature, art, drama, music), they are a plurality

> of *transdisciplinary, core-disciplinary, artistic practices, processes, and paradigms* that spill over, usually at the deepest point, into all disciplines.
> (2008, 231, italics in original)

Postdigital storytelling – hybridic, embodied, affective – becomes a powerful transdisciplinary research method that reaches out, or spills over, to use Johnston's phrase, into other, non-disciplinary areas. The translation of knowledge across these different domains and cultures is a central part of this process, forming new string-figures or meshworks, new kinds of interdependencies.

Yet ultimately, transdisciplinarity also offers something more profound: a way of engaging with global priorities that is not limited to previous approaches or ways of thinking. It is an approach that says, if we are to continue to survive amidst the mounting crises of the Anthropocene, then we need to think beyond the limits of academic boundaries and convention. In other words, we need to embrace new ways of making trouble.

Social empathy and resilience

A good example of this sort of global imperative that nevertheless connects strongly with the arts and humanities would be the emergence of what has been broadly termed social and cultural resilience. Resilience of course is not a new concept; in the sciences, it refers to the ability of an object or substance to retake its original shape after the application of an external force. When used in reference to people, or human organisations and structures, it refers to their capacity to overcome difficulty or adversity. This focus on the human implications of resilience has grown in importance over the last few years, driven, at least in part, by our own increasing awareness of mounting global crises. A lot of these crises are directly associated with the ecological impact of the Anthropocene, of course. Yet, as we've seen with the AHRC's Connected Communities Programme, for example, there is an important social and cultural dimension to any concept of human resilience that goes beyond a straightforward linkage with ecological degradation. By placing the creation of sustainable and resilient communities at its core, the Connected Communities Programme engages with the issues surrounding urban living, heritage, wellbeing and civic engagement across diverse and fragmented societies. As I've shown, postdigital storytelling should be seen as a key research methodology in this context, a form of knowledge making that prioritises the embodied materiality of its participants. Social and cultural resilience demands, by its very nature, a transdisciplinary approach to research in which the translation of knowledge across 'discursive, material and institutional boundaries' (Pahl and Facer 2017, 224) is foregrounded. It is this translation across, this striving for 'thinking differently' (2017, 226), that makes storytelling such a powerful methodology.

In *The Empathy Instinct: How to Create a More Civil Society*, Peter Bazalgette argues that cultivating empathy across the world has become one of the most pressing issues facing humankind. Without empathy there can be no common understanding, respect or love, no sense of commonality or connection, that makes possible a reaching out beyond narrow divides and entrenchment. Although he doesn't use the term directly, Bazalgette's book is all about social and cultural resilience. In his ten-point Charter for Empathy (2017, 292–305), Bazalgette prioritises the importance of, what he terms, the 'empathetic arts' (2017, 302), in other words, the kind of creative storytelling that has been at the heart of this book. For Bazalgette, the arts and humanities are critical in how we engage with the fundamental issues of what it means to be a human being in the twenty-first century. As he states:

> It's clear that if we ensure each generation immerses itself in arts and culture, in all its many manifestations, we'll build better citizens who understand each other's feelings and needs. This is what it is to be human.
> (2017, 290)

Elizabeth A. Segal, in *Social Empathy: The Art of Understanding Others*, makes a distinction between what she terms interpersonal and social empathy. The former describes a lower order of empathy, in which the emphasis is on empathetic expression between individuals or small groups (2018, 3). According to Segal, interpersonal empathy includes the ability to both mirror the physiological actions of another *and* take another's perspective while at the same time being aware that the experience is not one's own (2018, 3). Social empathy, on the other hand, is much broader and deeper as a concept: rather than being limited to individuals and small groups, it refers to any empathic understanding between large numbers of people across diverse social groups. For this to happen, Segal isolates two unique features which are not present during interpersonal empathy: the first is contextual understanding, and the second, macro perspective-taking (2018, 14). Contextual understanding comes through an engagement with the historical context of a social group; macro perspective-taking, on the other hand, involves trying to empathise with those who are different from us at a macro, or large group, level. As Segal states, '[t]he more we contextualize our knowledge and engage in macro perspective-taking, the deeper our understanding' (2018, 177).

Both interpersonal and social empathy offer a means by which social and cultural resilience can be augmented. Yet Bazalgette and Segal are only half right when they recognise the importance of storytelling as a key means by which such empathetic understanding can be deepened. What this book has shown is that postdigitality offers a fundamental reconceptualisation of creative practice that is itself posited within the broader retreat from postmodernity. From this critical perspective, the hybridic, embodied and affective storytelling that I've mapped out across the chapters of this book become key to understanding how empathic understanding might be strategically developed and encouraged. The sort of social empathy that Segal describes is a translation across boundaries and social division, the same sort of transdisciplinary enmeshment that lies at the heart of postdigitality. And just like empathy, at the heart of postdigitality lies the fundamental imperative to connect and understand. If Paglen's *Fanon (Even the Dead Are Not Safe) Eigenface* is an insight into the inscrutable gaze of AI, the ultimate digital-to-digital storytelling, then the stories I've discussed throughout this book offer an alternative way of thinking about our relationship with technology, a relationship that priorities our own humanity, encapsulated by the reproduction of Vija Celmins' work, *Concentric Bearings B*, at the very end of Ben Lerner's novel, *10:04*, and discussed in Chapter 6.

Post-postdigital

The end of a book is also the start of new beginnings, of new horizons of possibility. And the same is also true here. Postdigitality arose in the early years of the new millennium as a direct response to the notion of digital/non-digital binarity, a pulling back from the extreme technological determinism of the 1990s. In part driven by the DIY hacker ethic of artists, it was also given traction through the demise of postmodern scepticism and the rise of new forms of ethical sincerity and personal affectivity. I would suggest that these developments are not going to go away any time soon. Instead, what might emerge from our postdigital era is an increasing normalisation of hybridity in which any teleological understanding of technology, the progression towards some kind of cybernetic utopia, is stripped away, leaving instead a far-more open relationship, in which entrenched views of technology as non-digital and digital, low-fi and hi-fi, redundancy and cutting edge, become increasingly irrelevant. Yet, it is also clear that the urge to tell stories will not recede; indeed, this book would suggest that these changes will only add to the range and form of the stories we tell each other. Making sure it stays abreast of these social and cultural changes will continue to be a critical task for the arts and humanities.

Note

1 Rosemary Ross Johnston, 'Transdisciplinarity and the Arts', 225.

Works cited

Abba, T. and Speakman, D. *These Pages Fall Like Ash*. Bristol: Circumstance, 2013.

van den Akker, R. and Vermeulen, T. 'Periodising the 2000s, or, the Emergence of Metamodernism'. *Metamodernism: Historicity, Affect, and Depth After Postmodernism*. Eds, van den Akker, R. Gibbons, A. and Vermeulen, T. London: Rowman & Littlefield International, 2017, 1–19.

van den Akker, R. Gibbons, A. and Vermeulen, T., eds. *Metamodernism: Historicity, Affect, and Depth After Postmodernism*. London: Rowman & Littlefield International, 2017.

Arendt, H. *The Human Condition*, Chicago: University of Chicago Press, 1998.

Attlee, J. *The Cartographer's Confession*. James Atlee, 2018.
Bazalgette, P. *The Empathy Instinct: How to Create a More Civil Society*. London: John Murray, 2017.
Berry, D. M. and Dieter, M. 'Thinking Postdigital Aesthetics: Art, Computation and Design'. *Postdigital Aesthetics: Art, Computation and Design*. Eds, Berry, D. M. and Dieter, M. London: Palgrave Macmillan, 2015, 1–11.
Bolt, B. *Heidegger Reframed*. London: I.B. Tauris, 2011.
Braidotti, R. *The Posthuman*. Cambridge: Polity Press, 2013.
Casone, K. 'The Aesthetics of Failure; Post-Digital Tendencies in Contemporary Computer Music'. *Computer Music Journal*, 24 (4), 2000: 12–18.
Cornell, L. Bryan-Wilson, J. and Kholeif, O. *Trevor Paglen*. London: Phaidon Press, 2018.
Cramer, F. 'What is "Post-Digital"?'. *Postdigital Aesthetics: Art, Computation and Design*. Eds, Berry, D. M. and Dieter, M. London: Palgrave Macmillan, 2015, 12–26.
Danielewski, M. Z. *House of Leaves*. London: Transworld Publishers, 2000.
Facer, K. and Pahl, K. 'Introduction'. *Valuing Interdisciplinary Collaborative Research: Beyond Impact*. Eds, Facer, K. and Pahl, K. Bristol: Polity Press, 2017. 1–21.
Gillespie, W., Rettberg, S., Stratton, D. and Marquardt, F. The Unknown. 1998. Web. 30 April 2018. http://unknownhypertext.com.
Haraway, D. J. *Staying with the Trouble: Making Kin in the Chthulucene*. Durham: Duke University Press, 2016.
Holland, M. K. *Succeeding Postmodernism: Language and Humanism in Contemporary American Literature*. London: Bloomsbury, 2013.
Jackson, S. *Patchwork Girl; or a Modern Monster by Mary/Shelley and Herself*. Watertown, MA: Eastgate Systems, 1995.
Jackson, S. and Jackson, P. The Doll Games. 2001. Web. 28 May 2018. www.ineradicablestain.com/dollgames.
Johnston, R. R. 'Transdisciplinarity and the Arts'. *Transdisciplinarity: Theory and Practice*. Ed, Nicolescu, B. C., New Jersey: Hampton Press Inc.: 2008, 223–236.
Joyce, M. *afternoon: a story*. Watertown, MA: Eastgate Systems, 1987.
Kearney, R. *The Wake of Imagination*. London: Taylor and Francis, 1988.
Landow, G. P. Hypertext: The Convergence of Contemporary Critical Theory and Technology. Baltimore: John Hopkins University, 1992.
Lerner, B. *10:04*. London: Granta, 2014.
McIntosh, M. *theMystery.doc*. London: Grove Press, 2017.
Montuori, A. 'Foreword: Transdisciplinarity'. *Transdisciplinarity: Theory and Practice*. Ed, Nicolescu, B. C., New Jersey: Hampton Press Inc.: 2008, ix–xvii.
Nicolescu, B. '*In Vitro* and *In Vivo* Knowledge – Methodology of Transdisciplinarity'. *Transdisciplinarity: Theory and Practice*. Ed, Nicolescu, B. Cresskill, New Jersey: Hampton Press Inc.: 2008, 1–21.
Paglen, T. Shark, Linear Classifier. Dye Sublimation Print, 48 × 48 inches. Altman Siegel, San Francisco, 2017a.
Paglen, T. Shoshone Falls, Hough Transform; Haar. Silver Gelatin Print 48 × 60 inches. Collection of Rory and John Maxon Ackerly, 2017b.
Pahl, K. and Facer, K. 'Understanding Collaborative Research Practices: A Lexicon'. *Valuing Interdisciplinary Collaborative Research: Beyond Impact*. Eds, Facer, K. and Pahl, K., Bristol: Polity Press, 2017, 215–231.

Paul, C. and Levy, M. 'Genealogies of the New Aesthetic'. *Postdigital Aesthetics: Art, Computation and Design*. Eds, Berry, D. M. and Dieter, M. London: Palgrave Macmillan, 2015, 27–43.

Pope, R. *Creativity: Theory, History, Practice*. London: Routledge, 2005.

Pullinger, K. *Breathe*. Editions at Play, 2018. Web. 19 December 2018. https://editionsatplay.withgoogle.com/#/detail/free-breathe.

Pressman, J. '*House of Leaves*: Reading the Networked Novel'. *Studies in American Fiction*, 34 (1), 2006, 107–128.

Pynchon. T. *Gravity's Rainbow*. New York: Viking Press, 1973.

Ryan, M. L., Foote, K. and Azaryahu, M. *Narrating Space / Spatializing Narrative: Where Narrative Theory and Geography Meet*. Columbus: Ohio State University Press, 2016.

Segal, E. A. *Social Empathy: The Art of Understanding Others*. New York: Columbia University Press, 2018.

Smith, Z. *NW*. London: Hamish Hamilton, 2012.

Smyth, K., Power, A. and Martin, R. 'Culturally Mapping Legacies of Collaborative Heritage Projects'. *Valuing Interdisciplinary Collaborative Research: Beyond Impact*. Eds, Facer, K. and Pahl, K. Bristol: Polity Press, 2017. 191–213.

Speakman, D. *It Must Have Been Dark By Then*. Bristol: Taylor Brothers, 2017.

Tabbi, J. 'Introduction'. *The Bloomsbury Handbook of Electronic Literature*. Ed, Tabbi, J. London: Bloomsbury, 2018, 1–9.

Walsh, J. *Seed*. London: Editions At Play, 2017. Web. 5 February 2018. https://seed-story.com.

Index